THE GRIZZLIES OF MOUNT McKINLEY

For sale by the Superintendent of Documents, U.S. Government Printing Office
Washington, D.C. 20402

Adolph Murie on Muldrow Glacier, 1939.

THE GRIZZLIES OF MOUNT McKINLEY

Adolph Murie

Scientific Monograph Series No. 14

U.S. Department of the Interior
National Park Service
Washington, D.C.
1981

As the Nation's principal conservation agency, the Department of the Interior has responsibility for most of our nationally owned public lands and natural resources. This includes fostering the wisest use of our land and water resources, protecting our fish and wildlife, preserving the environmental and cultural values of our national parks and historical places, and providing for the enjoyment of life through outdoor recreation. The Department assesses our energy and mineral resources and works to assure that their development is in the best interests of all our people. The Department also has a major responsibility for American Indian reservation communities and for people who live in Island Territories under U.S. administration.

Library of Congress Cataloging in Publication Data
Murie, Adolph, 1899–
The grizzlies of Mount McKinley.

(Scientific monograph series; no. 14)
Supt. of Docs. no.: I 29.80:14
Bibliography: p.
Includes index.
1. Grizzly bear—Behavior. 2. Mammals—Behavior. 3. Mammals—Alaska—Mount McKinley National Park. 4. Mount McKinley National Park. I. Title. II. Series: United States. National Park Service. Scientific monograph series ; no. 14.
QL737.C27M87 599.74′446 80-607120

Contents

Figures

Tables

Preface

This monograph is based on observations of grizzly bears in Mount McKinley National Park* by my father, Adolph Murie. He spent a total of 25 summers in the park from 1922 to 1970. In earlier years information was obtained incidentally in the course of studies on wolves and ungulates; after the mid-1950s, he concentrated his efforts on grizzlies.

He spent long hours observing bears, even when the animals were engaged only in feeding on vegetation or resting. The development of an interesting event in the inter-relationships of bear families or of bears with other species was so unpredictable that he tried to be on the spot when it occurred. Often he followed a bear family for several consecutive days as it traveled through the park in a course parallel with, and visible from, the park highway. Because certain characteristics were apparent to him, he could distinguish quite accurately the different bear families, mothers, and cubs.

A number of people who traveled regularly in the park, such as photographers and park personnel, kept him informed on the locations of bears. He was careful to examine his sources of information for reliability and accuracy, and he drew conclusions from his observations with care.

At the time of his death in 1974, he had completed drafts of most sections of the manuscript and was in the process of incorporating observations from his last three summers in the field. I completed this task and have written several of the short sections which were sketched out only roughly. In this and in editing completed sections, I have tried to retain the spirit in which he wrote. However, I am sure, had he been able, he would have added considerably more polish to the manuscript. His approach to writing was literary, and is reflected, perhaps, in a quotation by Margary Allingham that he saved: "I write every paragraph four times: once to get my meaning down, once to put in everything I left out, once to take out everything that seems unnecessary, and once to make the whole thing sound as if I had only just thought of it." He also took the advice of his brother, O. J. Murie, who in 1962 wrote in a letter: "It seems to me we should get away from the strictly scientific methods of today, so much like the laboratory technique. We have to

*The name was changed to Denali National Park in December, 1980.

speak the truth but we can use human language in doing so.'' In an early draft of the introduction, Adolph wrote, ''I have, I think, avoided the ecologist's jargon, the scientific phrases so frequently created by ecologists and animal behaviorists to make simple facts sound profound and impressive.''

This monograph is a report on the behavior and ecology of grizzly bears in McKinley National Park. Adolph inserted a few references and I have added others, mainly where a comparison of quantitative results (litter size, density, etc.) seemed appropriate. However, it is obvious that he did not intend this to be a comprehensive monograph on grizzlies throughout their range.

Adolph held strong philosophical views about biological studies in national parks; some of these are apparent in the text. Although he recognized that studies of marked bears would yield additional data of value, he felt strongly that marked animals are out of place in national parks. It was his view that the aesthetic experiences possible in a wilderness park such as McKinley should be cherished, and National Park Service policy should work to promote such experiences. I think his attitude is well expressed in a quotation which he copied just before his death from *The Wilderness of Beauty* by Edward Graves: ''This perfection is much more likely to be realized where the hand of man is only reverently and lightly laid upon it.''

JAN O. MURIE

Acknowledgments

In the course of a 20-year study of the grizzly bears of Mount McKinley National Park, Adolph Murie had the continuous support of the National Park Service by whom he was employed during the major part of the time. Also during those years the personnel at the park were interested and helpful. Many of the Park Rangers reported their observations of grizzlies in the course of their regular work, and the bus drivers were also helpful in this regard, as they traveled the park highway daily. In addition, there were numerous wildlife photographers who recounted their experiences with grizzlies.

Special thanks are due to Charles Ott, a professional photographer who was employed at McKinley National Park during several years of the study and who was most generous with advice on camera cquipment and with his own photographs.

For plant identification the author had the superior knowledge of Dr. Eric Hulten, the foremost authority on arctic flora around the world, from the Naturhistoriska Riksmuseet in Stockholm, Sweden. Dr. Hulten spent 2 weeks with us in the field, collecting and pressing plants, and it was a distinct advantage to be able to discuss various species of plants with him.

Murie's son, Dr. Jan O. Murie of the Zoology Department at the University of Alberta, Canada, was responsible for organizing the vast amount of written material in publishable form.

Special thanks are extended to Dr. Allen Stokes of Utah State University at Logan, Utah, who reviewed the manuscript and suggested changes, many of which were incorporated, and to Dr. F. C. Dean of the University of Alaska for his review of a later draft.

I am grateful to Margaret Murie who assisted me in typing the manuscript.

Moose, Wyoming LOUISE G. MURIE

Summary

This report includes my observations on grizzly bears (*Ursus arctos* L.) in Mount McKinley National Park from 1922 to 1970; studies were most intensive from 1959 to 1970.

Grizzlies range throughout the park, but favor particular areas where food is abundant. Density in a 400 square mile area along the road where most work was done was estimated at one to two bears per 10 square miles. Mean litter size was 1.85 for spring cubs and 1.70 for all age-classes of cubs.

Home ranges were documented for 2, 3, or 4 years for a number of bears, primarily families that I recognized from year to year based on characteristics of females and cubs. Bears tended to occupy the same general area every year. Observed ranges, usually 5 to 12 miles in length and 1 to 5 miles wide, do not represent total home ranges because rough terrain limited observability. Bears occupy different portions of their home ranges as food availability and food habits shift from season to season. Home ranges overlap extensively and territoriality was not evident. A sort of "peck order" based on size, and perhaps reproductive status and past experience, determined the outcome of encounters between bears. Ordinarily, bears avoid close proximity to others.

The breeding season extends from mid-May to mid-July, with a peak in June. In spring, males wander widely in search of receptive females. A male attends one, or occasionally two, females for 1 to 3 weeks. Initially, females are intolerant of males, often attempting to evade their attentions, but later become tolerant and permit the male to mount. The minimum breeding interval for females is 3 years, but is usually at least 4 years. Presumably, cubs are born in January and February. They remain with their mother until 2½ years of age, continuing to nurse into the spring and summer of their third year. Occasionally, a single cub stays with its mother into its fourth summer of life. Breakup of the family usually was initiated by the mother. After separating, twin and triplet cubs often remain together, at least in loose association, for up to three summers.

Grizzly bears are omnivorous, but rely mainly on a vegetarian diet that changes as summer progresses. During May and early June, digging for roots is the predominant feeding activity. Bears graze on grasses and herbs in late June, July, and to some extent in early August. Berries become a major food in August, and rooting activities increase again, especially in years when berry crops are poor. In September, digging for roots and ground squirrels are the most frequent feeding activities. Carrion is eaten whenever available, and bears occasionally capture young calves of moose and caribou in early summer. A large carcass often attracts several bears, but the largest bear in the area has priority.

Grizzly bears are potential or actual predators on a number of mammals sharing their range. Caribou and moose are wary of bears during their calving periods when bears actively prey on newborn animals. Caribou calves soon mature enough to outrun grizzlies, and caribou herds then pay less attention to passing bears. Cow moose with calves are usually able to defend their offspring from bears. Dall sheep are not vulnerable to bear predation most of the time when in their usual rugged and rocky haunts. During short migrations across valleys from winter to summer ranges, ewes and lambs are more subject to predation; bears occasionally catch a sheep then, usually by surprise in gentle terrain.

Of the smaller mammals, only ground squirrels are captured routinely by grizzlies. Bears are always alert for opportunities to surprise a ground squirrel away from its burrow, and in the fall may concentrate on digging them out for days at a time. Marmots and beaver rarely are captured. Porcupines are well protected against bears; their quills can cause temporary lameness to imprudent bears.

Bears meet a variety of other animals at carrion. Magpies and ravens obtain a small share with little problem. Wolves, however, have little chance to feed at a carcass if a bear is present, but are able to take their turn after a bear has temporarily had his fill.

Wild grizzlies in McKinley National Park, conducting their affairs undisturbed, are the essence of wilderness spirit.

Fig. 1. Denali (Mt. McKinley) stands above the grizzly's domain in Mt. McKinley National Park.

1 Introduction

On our initial day in the field in McKinley National Park in 1922, my brother and I were crossing from Jenny Creek over a rise to Savage River on our way to the head of the river. In those days there was no road, the park was all a blessed wilderness, and I have often thought since what a wonderful people we would have been if we had wanted to keep it that way.

I had never seen a grizzly, and we did not see one on our 20-mile hike, although it was superb bear country. One lone track in a patch of mud is all we saw. In innocent wonder I gazed at the imprint. It was a symbol, more poetic than seeing the bear himself—a delicate and profound approach to the spirit of the Alaska wilderness. Since that time, I have spent many joyful days in McKinley National Park, and many of them were devoted to observing grizzlies and grizzly sign (Fig. 1).

The data recorded in this book were gathered over a long period in the park observing many species of birds, mammals, and plants. Sometimes the data gathered were incidental to other projects, but in later years I was able to devote more time to observing bears.

Because we are dealing here primarily with grizzlies in a national park it may be well to ask, "What is a national park, what are its objectives, and what should we seek to preserve?" Through the years there have been varying viewpoints. For instance, for a number of years the superintendents in Yellowstone National Park were interested chiefly in preserving ungulates such as elk, bighorn sheep, and deer; carnivores such as cougars, wolves, and coyotes were destroyed to that end, they thought.

In a 1963 report on wildlife management in the national parks, a special committee appointed by the Secretary of the Interior set forth the objective of national parks as follows: "As a primary goal, we would recommend that the biotic associations within each park be maintained or, where necessary, recreated, as nearly as possible in the conditions that prevailed when the area was first visited by the white man. A national park should represent a vignette of primitive America." Offhand, this statement has the ring of idealism. But it says we should freeze Nature and stabilize the environment, through management, at the stage when first seen by white man. It advises that man take charge and halt the natural ecological processes.

Another committee, under the aegis of the National Academy of Sciences, restated these objectives and returned them to what many of us feel is acceptable and the original objective in creating parks. For any habitat to have full significance we must try to maintain all the natural ecological factors and leave them as undisturbed as possible. In McKinley National Park man has an opportunity to be especially virtuous, and an obligation to come closer to the ideal than in more population-centered parks.

Much has been written about bears throughout history, but until recently we have known little of the detail of the natural history of grizzlies. It is always difficult to separate fact and fiction concerning an animal as awesome as the grizzly bear, but even in some of the old fables about bears one can extract some grains of understanding. There is an old fable, which amused the Eskimos when my brother told it to them, concerning how the bear lost his tail. Upon seeing a fox trotting along with a fine fish in his jaws, the bear entreated him to tell how one could obtain such a meal. The fox showed the bear how to hang his tail through a hole in the ice and, after it was frozen solid, told him to pull hard and he would have a nice fish. When the bear pulled, his tail came off and he has been essentially tailless ever since.

Legend has it that loss of most of his tail affected the bear more deeply than generally is suspected. It made him fat. When bears had long, bushy tails, they wrapped them around themselves and kept warm and snug in their hibernating caves during the cold winter months. The loss of the tail created a survival problem that was solved by building up a thick layer of fat under the hide. To build up this layer of fat the bear had to eat great quantities of food all summer. He had to begin in the spring when the first edible food consisted of berries, that had been frozen all winter, and roots. He had to eat anything and everything and became omnivorous. He ate great quantities of grass and its bulk made his stomach hang low; when berries came, he ate them all day long and in the fall went back to roots again just before hibernation. He was a carnivore; he loved meat, but over much of the land he could never get as much as he wanted because his stomach was so full he could not run fast enough to catch big animals and little ones would not fill him up enough. However, he managed to dig out an occasional ground squirrel and had to fill up on vegetation. Of course, some land treated him better than others, and where salmon spawned he feasted on them. Although the loss of the tail made the bear fat, it first made him a big eater. He had to eat so fast to get enough that he also lost his manners, and gobbled berries steadily, leaves and all. When he became fat enough to keep warm all winter he became so heavy that his legs had to grow big and strong. The loss of his tail made his temper uncertain and he became very temperamental and sometimes dangerous. When he was angry, he had not enough tail to take up the excess energy in slow writhing, as do cats and the energy then went into his legs and he charged toward whatever annoyed him. And he had not tail enough to put between his legs and run away, like a dog.

So the fox changed the bear a great deal more than is at first evident.

Such legendary flights of fancy are not so very different from impressions that people often gain regarding grizzlies. It is easy to misinterpret many aspects of grizzly behavior, particularly when confronted by a bear at close quarters and any forward movement becomes a "charge" in one's mind. On occasion, I have seen and heard of grizzlies walking along, so oblivious to any presence that one might suspect they were blind or that they were advancing toward one with malicious intent. In an article about his many experiences with bears, Earl Fleming (1958) attempted to debunk some of the myths that have grown up about them. He believes that "most bears accused of charging were not actually charging at all"—and I think he is right. He concludes that men confronted by bears seldom underestimate their number and size, the size of their tracks, or the danger to themselves. Much of the mystique surrounding grizzlies may never be dispelled, and perhaps that is good, as long as we maintain a reverence for the continued existence of bears and preserve areas such as McKinley National Park in such a way that they may continue to live without harassment by man.

Earl Fleming concludes his article with these words: "It would be fitting, I think, if among the last man-made tracks on earth could be found the huge footprints of the great brown bear."

2
Study Background

Classification and Characteristics

Taxonomy of Grizzly and Brown Bears

C. Hart Merriam (1918) published a classification of the grizzly and brown bears. For several years he had been gathering material, chiefly skulls, from hunters and others. He recognized 86 forms, most of them full species. Mammalogists early questioned the validity of the many species, most of which Merriam himself described. Judgments in the field of taxonomy are questioned frequently but usually not to the extent that the grizzly species were. It was believed that Merriam's entire grizzly classification was based on false premises, that he had assumed wrongly that variation found in skulls represented species rather than types of individual variation, such as we find in humans.

In the early 1930s my brother Olaus and I had the pleasure of spending an evening with Dr. Merriam in his Washington, D.C., home. He was a most colorful individual and an outstanding raconteur. He regaled us with stories of his biological explorations in the West, from as far back as the 1870s. He also told of an incident that concerned the dispute surrounding grizzly taxonomy and Theodore Roosevelt, who was among the doubters of the many grizzly species recognized. The incident took place at the Cosmos Club during a meeting to which President Roosevelt was invited. Dr. Merriam arrived early, carrying two grizzly skulls which he placed on the mantelpiece. Roosevelt soon spied the skulls, and a lively discussion followed. He conceded that the skulls were different enough to represent two species. Then followed Dr. Merriam's triumph—he told the President that the skulls represented two species which he had questioned. Merriam thus clinched his argument about his grizzly taxonomy and no doubt added to his certainty concerning the validity of his many species. My brother and I were sorry that we were still among the doubters.

Robert Rausch (1953), who has discussed bear taxonomy in Alaska, found wide variations in a number of skulls (22 with full data) gathered in the Brooks Range. He concluded that he was dealing with individual variation in an interbreeding bear population. He and others have, in

Fig. 2. The shoulder hump and dished facial profile easily distinguish grizzlies from black bears in the field.

recent years, placed the North American grizzlies and brown bears, along with the Eurasian brown bears, in a single species, *Ursus arctos*. Rausch lumps about 14 species listed for northern and central Alaska into one subspecies, *Ursus arctos horribilis*. Following Rausch's classification, the grizzly in McKinley National Park would be called *Ursus arctos horribilis*.

Description

In the field a grizzly bear may be distinguished readily from the black bear because of its pronounced shoulder hump. Also its facial profile differs from the straight profile of the black bear in being somewhat dished, that is, the forehead tends to rise and give a break in the profile line (Fig. 2). Black bears are generally black or brown, while grizzlies show a wide range of intermediate hues and have a grizzling over much of the pelage.

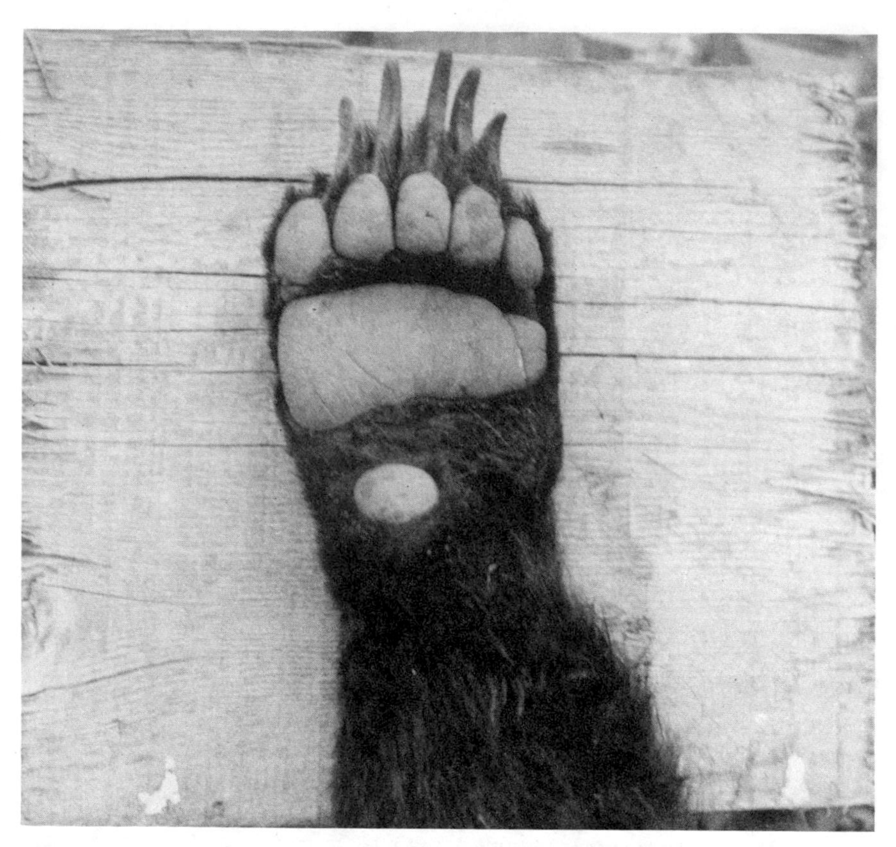

Figs. 3,4. Forefeet (above) and hind feet (right page) of grizzlies are clearly distinguishable. The long claws on the forefeet show in most tracks and contrast with the shorter, more curved claws of black bears.

In contrast to black bears, which have rather short and very curved claws on the forefeet, grizzlies have very long claws on the forefeet, more than 2 inches long unless badly worn, and only slightly curved. The middle foreclaws of an adult female measured 3½ inches along the dorsal curve and 2½ inches in a straight line from the base to the tip. Digging for roots and for ground squirrels tends to wear away the tips. The claws on the hind feet are much shorter and curved more sharply. The color of the claws in grizzlies can vary from dark brown to almost white (Figs. 3, 4).

The relatively straight foreclaws are not suited for climbing tree trunks. I have seen spring cubs and yearlings climb 10 or 12 feet from the ground in willows, clambering about in play. Ordinarily, adult grizzlies do not engage in tree climbing even in areas where trees are plentiful.

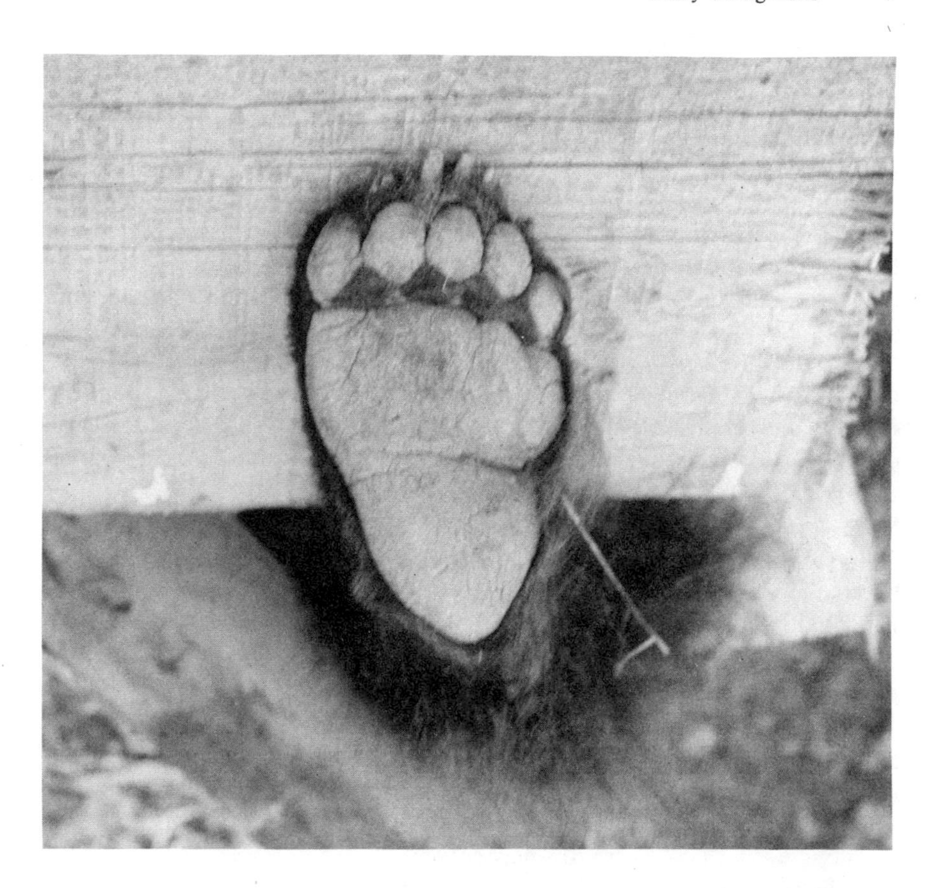

Grizzly bears have up to 42 teeth. One or more of the small premolars may be missing. The molar teeth are considerably flattened, apparently an adaptation to the omnivorous diet. However, the mastication of the vegetation eaten is very slight, as is shown by the remains found in the droppings.

The majority of the grizzlies in the park are light tan over most of the body, often referred to as blond. Head, neck, and shoulders may be light, almost buffy white, with legs and belly dark brown (in fall fur is shorter and darker.) Some are reddish-brown, some a rich dark chocolate, and a few almost black. An old male I examined was black except for dark-brown grizzling over the shoulders and back. The feet, legs, and underparts of the body are dark in all color phases, varying from blackish to various shades of dark reddish-brown. The face is generally slightly darker than the body but an occasional bear has such a light face that the dark eyes show prominently. The pelage is usually grizzled, the tips of the hairs being light. This is not evident in many dark bears. The shade of color of a bear varies according to the direction from which the light strikes it, relative to the position of the viewer. The bear appears darker when facing away from the light because of reduced reflection. On one occasion a tourist told me that he had seen a bear on Sable Pass that was light on one side and black on the other. I had noticed this striking difference in this particular bear too. Both sides were alike in color but as the bear shifted its position in relation to the light, the color tended to vary from blondish to blackish. The effect of light direction on the color of this blondish bear was more extreme than noted in any other bear. Possibly this was due to some special character of the grizzling in this individual.

The large grizzlies of the Alaska coast and adjacent islands, generally called brown bears, are colored more uniformly than bears in the interior, and usually are dark brown, although occasionally light-colored ones are reported in coastal populations.

Pelage color usually undergoes seasonal change. The grizzlies in McKinley National Park emerge from hibernation with their autumn coats still in excellent condition. As spring progresses, the fur tends to fade and lighten in color. The long northern days and light reflection from snowfields probably accelerates this fading. By July the fur on a few bears becomes somewhat ragged in a patch or two, but in most bears the changing of coat is hardly perceptible.

In my notes over the years I find references to indications of shedding in some bears. These observations are concentrated in the month of July, some in August, and one year I noted the shedding of a 4-year-old was still not completed on 18 September. Usually by September the bears, with few exceptions, have new coats. They are much darker than the old coats and appear rich and alive.

The spring cubs are blackish or dark brown in their spring coats. Some have a white vertical streak on the sides of the neck which usually is lost by the time they are 2 years old. By August they have acquired a new coat and show a lighter grizzling that is more pronounced in those cubs that were brownish in the spring.

Little information is available on weights of grizzlies in McKinley National Park. An old male grizzly that was not very fat weighed 650 pounds. The animal had been shot so there had been a loss of blood which could not be calculated. There was no opportunity to weigh a female but it is likely that females weigh about 200 pounds less than males. The difference in size shows up strikingly when one observes a mated pair.

The measurements of an old male and an old female from the park are as follows:

Sex	Date	Total length inches	Tail inches	Hind foot length inches	Front foot width inches	Weight pounds
Male	11 Sept. 1951	73	7	11	6.4	650
Female	28 Aug. 1963	56		9	5.25	

The difference in size between the sexes also is indicated clearly in skull measurements. A male skull, 15.8 inches long, weighed 4 pounds 12 ounces; a female skull, 13 inches long, weighed 2 pounds 7 ounces.

In the Brooks Range, Dr. Robert Rausch found grizzlies weighing from 400 to 700 pounds. The range in size of adult bears in the park is apparently similar to that from the above locality. The coastal and island grizzlies are much larger and are known to attain a weight of at least 1,200 pounds. Estimates run even higher. It is thought that coastal bears attain their large size because of the abundance of protein foods, mostly fish.

Age and Mortality

In zoos, grizzlies have lived for almost 30 years. That some bears in the wild live to a ripe old age is shown by the thorough wear we find on the teeth. One old male grizzly showed excessive tooth wear: four of the molariform teeth were worn in two, only two root stubs remaining in each, and one molar was missing; the upper and lower incisors were worn to the gums; the two upper canines were worn but still retained their shape, but the two lower canines were worn until only blunt stubs remained. This old male, shot as he was breaking into a work camp after several raids, had lived long enough to have worn out his teeth.

A female, mother of three spring cubs, and apparently killed by another bear, was quite old. The teeth were not as worn as those of the male described above, but the incisors and molariform teeth were worn to the gums. The canines showed much wear and were quite blunted.

Because grizzlies are relatively scarce, it is seldom that one finds bone remains or a carcass, so little was learned of relative mortality rates in different age groups. Nevertheless, it may be of interest to list skull remains and carcasses found in the field.

Skull remains:

1. Mandible of spring cub.
2. Mandible of what appears to be a yearling.
3. Part of skull of young adult bear.

4. Skull of old female with some necrosis at base of a lower molar.

Carcasses found:

1. Two spring cubs killed by another female.
2. Very old female, mother of three spring cubs, at garbage dump. She was killed apparently by another bear.
3. A young adult female apparently killed by another grizzly.
4. Adult bear—cause of death not known.

From what little evidence is available, it appears that death at the hands of another grizzly may be a large part of mortality. Sufficient information to assess the effects of disease and parasites is lacking.

Use of Senses

Like many other mammals, grizzlies rely extensively on their sense of smell in conducting their day-to-day activities, although sight and hearing also play a role. Initial awareness of the presence of potential prey, other bears, or possible competition for carrion such as wolves or wolverines usually seems to be accomplished by detecting their scent.

I have watched bears, nose to ground, move about as though following a trail in areas being traversed by migrating caribou. Once a bear behaved in this manner for several minutes, eventually flushing out a female caribou with a young calf and successfully capturing the calf. It seemed to know by the scent that a caribou calf was in the area.

On some occasions when a grizzly is concentrating on ground squirrels, it may hear one calling or see it, gallop to the spot where the squirrel enters a burrow, and begin to dig it out. The squirrel sometimes escapes from another exit, unnoticed, while the bear concentrates on digging. Although the escape was not observed, the bear soon ceases in his efforts, realizing, by the lack of ground squirrel odor presumably, that his quarry has left. Then he either moves on or follows the trail of the squirrel to another burrow and exerts himself again. In one instance, a bear followed the escapee's trail for about 100 feet and was rewarded for the effort by capturing the unlucky squirrel. More often, an escaping ground squirrel is seen by the bear and pursued immediately to the next burrow.

Bears are reported to have relatively poor vision, at least at long distances. My observations do not contradict this. Individual visual recognition, within families for instance, sometimes appears to be unreliable, especially if cubs become separated from mothers by several hundred yards. At such times there can be much hesitation on the part of the cub to rejoin the mother, even though they are in sight of each other. Olfactory reassurance that an adult is indeed its mother seems prerequisite for a cub to resume its usual activity within the family.

The use of hearing by bears is not as obvious as that of sight and smell. Cubs do respond at some distance to low "woofs" or grunts from the mother; and I have noted that bears can detect slight noises that I have made at distances of 200 yards. Even though it may not play a prominent role in their activities, I believe grizzlies do have an acute sense of hearing.

Habitat

Much of McKinley National Park is treeless tundra, but strips of woods follow the rivers far into the park, and patches of trees grow here and there on adjacent mountain slopes. Timberline varies according to soil and exposure; in places it reaches elevations of over 3,500 feet.

White spruce is the common conifer. Black spruce is confined to poorly drained and boggy areas. Along the north boundary I have seen a few patches of tamarack. Cottonwood and aspen are distributed widely and a few birches grow at lower elevations. Along the McKinley River an extensive strip of cottonwoods may be seen from the highway.

The tundra supports a growth of willow and dwarf birch. Over 20 kinds of willow occur in the park. They range in size from small forms only 2 or 3 inches in height to brushy growths 20 feet tall. In places, the small willows may grow dense enough to form a sod. These shrubs are highly important for wildlife. Alder brush is distributed widely and is plentiful on canyon slopes; near Wonder Lake there are many clumps of alder in the rolling tundra.

The low ground cover over the park consists of mosses, lichens, sedges, grasses, horsetails, and herbaceous plants—many species of each. Early flowers may begin to bloom in late April and early May, and at the higher elevations some blooms may be seen in later summer.

All of McKinley National Park can be considered bear country, except for the snow-covered upper reaches of the peaks of the Alaska Range. One may meet grizzlies anywhere, from river bars to ridge tops. Particular habitats used by bears vary with the season and from year to year, depending on food availability. In the spring, the river bars and some hillsides are favored places for digging roots of peavine and other plants. As green vegetation becomes available, many bears move to areas with grassy swales such as on Sable Pass where grazing, primarily on grass, becomes a principal activity. Berry crops appear later in the summer and the location of bear activity coincides with areas where blueberries, crowberries, or buffaloberries are abundant. In some years when berry crops are poor, bears wander more widely than usual in late summer. The habitat is sufficiently varied over most of the park that bears may find spring, summer, and autumn foods within a limited area.

Thus, bear habitat is affected by the vagaries of weather and its effects on the phenology of plant foods of bears. Another direct influence on bear habitat, though one that operates over the long term, is the change wrought by rivers on river bars. The many rivers in the park, such as

Fig. 5. Here a glacial stream has shifted its channel and is washing away an old vegetated river bar where bears, spring and fall, feed on roots. Old river bars are continually washing away and new ones forming.

the Sanctuary, Teklanika, East Fork, and Toklat, break up into numerous channels which are constantly shifting over the broad gravel bars. During the summer, they carry much glacial silt that is picked up in one place and deposited in another. Where the silt is deposited, the bottom of the channel may build up until the stream breaks over the edge to flow off to one side over a slightly lower part of the bar. Heavy rains greatly increase the volume, sometimes forming one sheet of water covering the entire river bed and causing many changes in the channels. By moving back and forth, the streams tend to keep widening the bars. Some bars are a half-mile or more in width (Fig. 5). Changes may be slow or rapid.

Over long stretches of the rivers, gravel bars have remained undisturbed long enough to become covered with a thin, firm sod. These old bars are delightful for hiking as they are covered chiefly with low-growing vegetation and are as smooth as a lawn. The grizzlies also find them delightful and spend much time in spring and fall digging the roots of the peavine which prospers in this habitat. Up toward the heads of the rivers another species of the pea family (*Oxytropis viscida*) flourishes and attracts the grizzlies who come to graze on its stems, leaves, and

Fig. 6. Here the glacial stream is washing away a wooded flat, thus extending the width of the bar and creating more peavine habitat for bears in the future.

flowers. Some of the old bars support good stands of buffaloberry, and these berries are an important food for bears. Thus it is apparent that the river bars represent a significant part of bear habitat.

These old river bars originate through the activity of the rivers, but the river may also destroy them. An old bar that has been left in repose for 50 or more years may suddenly be invaded by a shift of some of the stream branches. In the mid-1950s, part of the Toklat River swung westward and flowed over an extensive old bar, cutting new channels which braided and widened. This had been a favorite rooting area for the grizzlies. Similarly, along the East Fork River a channel swung sharply into an old, high river bar, supporting a good stand of buffaloberry, and reduced the size of this bar considerably.

In places, river bars may be left undisturbed long enough for choice grizzly habitat to disappear because of plant succession. Along the Teklanika River an old river bar has supported a good stand of peavine for years. Now a young spruce forest is taking over and is slowly causing a decrease in the peavine. In time, the peavine will be shaded out entirely unless the stream invades.

The duration of the cycle involving the change from gravel bar to vegetation-covered bar and back to gravel bar varies greatly. In some places, plant life is washed away before it has had time to form a good sod. In other, limited areas, a spruce forest has had time to develop before the stream has begun to erode, invade, and return the wooded area to a barren gravel bar (Fig. 6). Thus the cycle tends to keep the grizzly habitat along the rivers in balance.

Other habitats in the park frequented by grizzlies are not subject to changes of this sort. The progress of plant succession in some areas is very gradual in this northern climate, and long-term weathering processes occur, but most parts of the bear's domain are not altered drastically over the years.

Numbers and Density

Total numbers of bears in the park and changes from year to year are difficult to determine. Several factors contribute to this difficulty. One is size of area and rugged topography which precludes consistent sightings of bears, even in portions of the park visible from the road. The majority of effort was in areas visible from the road; year-to-year variations in season and relative abundance of various food sources resulted in different patterns of use by bears. Thus, in some years bears were concentrated in areas where sightings were relatively easy, whereas in other years bears were absent from these areas, resulting in fewer sightings. In addition, my intensity of effort varied from year to year, and prior to about 1959 there was no special concentration on grizzlies.

Few aerial counts of bears have been made in the park, but aerial counts probably are not as complete as ground counts anyway. In 1969, a pilot surveyed the park for bears and remarked on the scarcity of females with cubs; he saw mainly lone bears. This is the year in which I recorded 20 families, more than in any other year. Other bear researchers have remarked on the difficulty of spotting grizzlies from the air even when a bear has a radio collar and its general location is known (Herrero 1972:82).

For each year I have calculated minimum numbers of different bears seen. These figures are probably fairly accurate for families because variations in pelage characteristics of the cubs and females make individual identification possible in most cases. Numbers of lone bears, however, probably are substantially underestimated. I have calculated numbers conservatively by counting only those lone bears and families that I am confident are different from other sightings. These data are presented in Table 1.

Table 1. Minimum numbers of grizzly bears observed in Mt. McKinley National Park, 1939–1970.

Year	Size of litter			Females with cubs	Lone bears	Total adult bears	Total bears
	1	2	3				
1970	7	4	0	11	10	21	36
1969	12	8	0	20	18	38	66
1967	5	6	2	13	17	30	53
1966	5	4	1	10	15	25	41
1965	3	5	0	8	15	23	36
1964	5	8	0	13	28	41	62
1963	9	9	1	19	24	43	73
1962	6	9	0	15	29	44	68
1961	5	10	0	15	36	51	76
1960	4	13	0	17	22	39	69
1959	5	13	1	19	23	42	77
1956	5	6	1	12	19	31	51
1955	4	6	1	11	17	28	47
1953	3	2	2	7	13	20	33
1951	2	5	0	7	7	14	26
1950	0	1	1	2	11	13	18
1949	2	6	1	9	17	26	43
1948	3	5	1	9	12	21	37
1947	4	5	0	9	17	26	40
1945	1	0	0	1	9	10	11
1941	2	1	2	5	11	16	26
1940	1	3	3	7	14	21	37
1939	4	3	2	9	13	22	38

Numbers of different bears seen were particularly low in 1965, 1966, and 1970. There was no apparent reason for low numbers recorded in 1965. In 1966, spring was very late in the park and bears were not abundant in areas usually favored by them in other years. The information from 1970 is incomplete because I was in the park only from late May through June.

Local densities of bears within the park are difficult to calculate because bears move around considerably during the summer as food habits and food availability change. Greatest densities occur on Sable Pass. In 1961 and 1962, there were as many as 3½ to 4½ bears per square mile. A rough figure of density for the portion of the park that my observations covered can be arrived at by using an area 5 miles wide along the length of the road from park headquarters to Wonder Lake, a distance of about 80 miles. Densities in this 400 square mile area in the years from 1959

Fig. 7. A female grizzly with two spring cubs on Sable Pass, an area of the park favored by bears in summer for grazing.

to 1970 ranged from 0.9 to 1.9 bears per 10 square miles or, if cubs are omitted and only lone bears and family units are used, 0.5 to 1.3 bears per 10 square miles.

Pearson (1972) estimates a density of one grizzly per 10 square miles in Yukon Territory and Kistchinski (1972) suggests densities in northeast Siberia of from 1½ to 2½ bears per 10 square miles. In Glacier National Park, Montana, Martinka (1974) estimated a density of 1 grizzly per 8.2 square miles. Densities may be much higher, comparable to those recorded on Sable Pass, in small areas on the southwest coast of Alaska and in northeast Siberia where bears congregate to take advantage of a particular food source. Thus, density of grizzly bears in McKinley National Park is not dissimilar to that in other areas (Fig. 7).

My information on production of young also shows low production in the same 3 years that numbers were low, probably a result of bears being less observable in those years rather than actual lower recruitment rates. There seems to be no strong relationship between number of spring litters and number of litters of older cubs. Additional information on family statistics and breeding interval is presented later.

Age Determination of Cubs

In these McKinley National Park studies, I have separated cubs into three categories: first year or spring cubs, yearlings, and 2-year olds. In a few instances, cubs 3 and 4 years old were recognized because of my earlier acquaintance with them.

Spring Cubs

Data from bears living in zoos indicate that cubs are born in January and February after a gestation period of about 7 months. Seton (1929) writes that a newborn grizzly was only 8½ inches long, had a grizzly shoulder hump, a tail proportionately longer than that of an adult, and weighed 1½ pounds. It appeared to be naked but was covered with fine, short gray hair. In 1 hour and 40 minutes it began to nurse a foster-mother dog.

When the tiny cubs are observed abroad in spring and early summer, they still are surprisingly small and scrawny. They seem too tiny to be bears (Fig. 8). Their color is blackish, but on close look some are dark brown. These latter apparently become lighter, blondish bears as they age. Some cubs have a white vertical streak on one or both sides of the neck. The amount of white may vary from a thin line to a rather extensive patch, and can be used to identify the individual reliably (Fig. 9). Growth of cubs during the summer is slow. They become more roly-poly, and the fur becomes grizzled by fall. There is no difficulty in recognizing the spring cubs throughout the summer.

Yearlings

The yearlings in spring are about the size of the spring cubs in autumn. They are obviously not spring cubs and are too small for 2-year-old cubs, although even an experienced hunter may confuse them at times. Judging from field observations, there is sometimes considerable variation in the size of cubs in different litters in the same age categories. I recall a family of two yearlings that I knew as spring cubs that were especially small. A rather experienced bear-hunting guide thought they were spring cubs. When these cubs were 2 years old, they were still small and seemed too small for 2-year-old cubs. But their age was always determinable. A variation in size of cubs is sometimes shown strikingly in a single litter where one cub may be much larger than the other. Toward autumn, yearlings seem to be the size of spring 2-year-old cubs. At this time one might wonder occasionally whether cubs are yearlings or 2-year-olds if the yearlings happen to be especially large (Fig. 10).

Two-Year-Olds

Two-year-old cubs show considerable variation in size. I have seen some suckling that seemed too large for this age. Perhaps they were

Fig. 8. Mother with spring cub. The whitish strip on the cub's neck still shows in September.

Fig. 9. Some spring cubs have distinct white patches on the sides of the neck which usually disappear by the time they are 2 years old.

Fig. 10. Mother with yearling (same family as Fig. 8) in spring of 1964.

males and their birthdays fell in the early part of the parturition period. Small two-year-olds, if still with the mother by late summer, could perhaps seem small enough for large yearlings if one were not familiar with them. Families in these categories may be somewhat puzzling to the observer when seen for the first time (Fig. 11).

In the field, I frequently have noted that the size of cubs compared to their mother seems to vary a great deal from day to day. At times the cubs seem large, then again, small. I also have noted in examining several pictures of a family that the cubs seem large in some pictures and small in others. On the whole, with experience in observation, one usually can be quite sure of age determinations in the three age categories described here. On five occasions I saw a 3-year-old cub in the spring still with its mother. If I had not known the family, I would have assumed the cub to be a 2-year-old. After seeing this cub still with its mother, it occurred to me that on one or two other occasions cubs were 3-year-olds rather than 2-year-olds as I had assumed.

After the cubs have left their mothers, size is difficult to determine because there is no good method of comparison. A 2- or 3-year-old cub seen alone might be taken for an older bear. The body seems shorter in young bears but this is an uncertain criterion. Once I saw two 2-year-old cubs near other older cubs and could easily recognize them as smaller. But when I saw these two by themselves, their smaller size was not obvious.

Fig. 11. Mother with 2-year-old cubs feeding on first green grass blades (June 23, 1963).

It has always seemed to me that the closer one approaches a bear, especially a cub, the smaller he appears to be. (If danger is involved, of course, the opposite may be true.) I have been close to known 2-year-old cubs that seemed to be the size of yearlings. On one occasion, the remarks of a friend of mine were significant in regard to judgment of size. The mother of three spring cubs that had been visiting a garbage dump was found dead. When my friend, who had often seen the mother alive, saw the carcass, he exclaimed that this could not be the mother of the three cubs he knew because that mother was "huge" and this dead one was small. Yet her dark color and the presence of the orphans made identification certain. Incidentally, several of the photographers in the park became quite expert in recognizing ages of cubs and also in recognizing the different families. But, as a final word, size in big country is always deceptive.

Some Family Statistics

The usual number of cubs in a litter varies from one to three. An old-timer reported that he once saw a litter of four cubs in McKinley National Park; in other areas also, four cubs have been reported and litters of four have been recorded in zoos. Over a period of years I have recorded the number of cubs in 249 families for most of which the ages were known (Table 2).

Table 2. Frequency of litter sizes at different ages observed in grizzly bears in Mt. McKinley National Park.

	1-cub-family	2-cub family	3-cub family	Totals	Mean litter size
Spring cubs	21	36	11	68	1.85
Yearlings	25	41	5	71	1.72
Two-year-olds	30	38	1	69	1.58
Three-year-olds	5	1	0	6	1.14
Unknown age	14	19	2	35	1.66
Total	95	135	19	249	1.70

In a study of the Kodiak bears at a salmon stream, Troyer and Hensel (1964) found that 51% of 39 spring-cub litters contained three cubs, 26%, 2 cubs, and 23%, 1 cub. The mean size of the spring-cub litters of these bears was 2.36. In McKinley Park, the mean litter size of the spring cubs was only 1.85. The high protein fish diet of the Kodiak bears possibly accounts in part for the large litters, and the longer season and more favorable climate also may be factors. In other studies of grizzlies, mean litter sizes have been reported as follows: 2.12 in Glacier National Park, British Columbia (Mundy 1963); 2.19 in the Alaska Peninsula (Lentfer 1966); 1.58 in Kluane National Park, Yukon Territory (Pearson 1972); 2.2 in Yellowstone National Park (Craighead and Craighead 1967); and 1.7 in Glacier National Park, Montana (Martinka 1974).

Table 3 shows the number of spring cubs, yearlings, 2-year-olds, and 3-year-old cubs seen in different years. The observations for the different years are not all comparable. During many of the early years, my studies were concentrated on other species which cut down on the observations of bears, and in some later years I was not in the park all summer so observations were less intensive. The years most comparable are from 1959 to 1969. One cannot assess cub losses very accurately from the table because some intact families may have been present but not seen.

I obtain a crude assessment of cub loss by comparing numbers of spring cubs in each of the years 1959 to 1966 to the numbers of yearlings the following year, and to 2-year-olds 2 years later. Loss of spring cubs was 31%, whereas that of yearlings was 17%. These rates of loss are similar to those reported on the Alaskan peninsula (A. W. Stokes, pers. comm.) and in Yellowstone National Park (Craighead and Craighead 1967). Note the decline of families of three cubs with their age: 11 of spring cubs, 5 of yearlings, and 1 of 2-year-olds. This suggests that at least one cub from families of triplets is lost frequently. The increase of one-cub families with their age (21, 25, and 30 for spring, yearling, and 2-year-old cubs, respectively) also suggests loss from litters of two and three cubs.

Table 3. Minimum numbers of families of one, two, and three cubs observed in Mt. McKinley Park, 1922–1970.

	1922	1923	1939	1940	1941	1945	1947	1948	1949	1950	1951	1953	1955	1956	1959	1960	1961	1962	1963	1964	1965	1966	1967	1969	1970	
Spring cubs																										
1 cub			1	1		1		1			1	1	1		1	1		2	5	2			2	1		21
2 cubs	1		1							1	2		2		2	7	3	2	2	4	3		1	2	3	36
3 cubs			1	2					1	1		1	1					1	1				2			11
																										Total 68
Yearlings																										
1 cub			2		1		1	1	1			1	2	2			4	1	1	2	1		1	3	1	25
2 cubs		1		2			2	2	1		2	1	3	3	6	2	3	4	3	1	1	2		2		41
3 cubs			1		2									1	1											5
																										Total 71
Two-year-olds																										
1 cub			1				2		1			1	1	2	3	2		3	1	1	1	3		5	3	30
2 cubs		1						3	5			1	1	3	4	4	2	3	3	2		1	4	1		38
3 cubs				1																						1
																										Total 69
Three-year-olds																										
1 cub																			1					1	2	4
2 cubs																							1			1
3 cubs																										Total 5
Age unknown																										
1 cub			1		1		2	1			1				1	1					1	1	1	2	1	14
2 cubs			2	1	1		3				1				1		2		1	1	1	1		3	1	19
3 cubs								1				1														2
																										Total 35
																										Total 248[a]

[a]One hundred ninety-nine different families.

Tracks and Trails

A few years ago, planners for the National Park Service suggested building trails for humans up some of the rivers and elsewhere in McKinley National Park. The grizzlies could have told them what park personnel learned later: that one can travel with relative ease over most of the park without trails.

Generally, grizzlies travel about at random. The few short pieces of bear trail that I have noted in the park have been along streams bordered by spruces, in stretches where bears frequently travel because of the terrain.

These bear trails, on the hard ground, remind me of Thoreau's trail at Walden in deep snow: ". . . For a week of even weather I took exactly the same number of steps, and of the same length, coming and going, stepping deliberately and with the precision of a pair of dividers in my own deep tracks. . . ." For some reason the bears tend to step in the same tracks until, over the years, a series of depressions is worn. In a short stretch of trail along the Teklanika River, where the water washed against the spruce-grown banks, the track depressions on the firm ground were worn an inch or more in depth, and were roughly 10 inches wide and a dozen inches in length. In these trails the front and hindfeet had stepped in the same depressions. The distance between them ranged from 23 to 30 inches. On a few occasions I have watched bears step in old tracks crossing snowfields, not missing a track. One bear followed a track in the snow as it walked to the top of a slope, stepping in the old tracks, then on reaching the top, he turned around and came down the slope carefully stepping in the tracks again.

In a slow, walking gait the hind foot may fall in the track of the forefoot or behind it, but usually in walking the hind foot registers ahead of the forefoot on the same side. In galloping, as we would expect, the tracks of the hind feet register ahead of the forefeet tracks in each set of four tracks. The pattern varies. Both hind feet may strike anterior to both forefeet, but sometimes one of the hind feet may be opposite the anterior forefoot track. In one set I noted that the tracks in each jump formed a diagonal line; the trail consisted of a series of these diagonal lines.

Bears show a great deal of variability in the way they move, but large males generally have a distinctive, ponderous walk, seemingly less flexible than females and younger individuals. Although grizzlies appear slow and somewhat ungainly when ambling along and feeding, they are well known to be capable of rapid bursts when galloping. On one occasion a galloping bear seemed to get most of its power from the front legs, the hind legs being brought forward without pushing. The forefeet landed well apart and the hind feet came down close together. But this was a subjective impression.

In spring and early summer, when most of the ground is free of snow, bear trails may be seen crossing the many snowfields lying in the draws of mountain slopes and in the hollows out on the rolling tundra. Some snowfields are crossed because they happen to lie in the line of travel, but I have the impression that bears prefer walking on the hard snow. On 3 June I observed a bear digging roots on a steep hillside. He started down the slope at a walk, catching himself at each step because of its steepness. I suggested to my companion that the bear could descend more easily if he would use the snowfield about forty yards to his right. A moment later he did just that and found travel easier. At first he sank in the snow with each step, but there was no jolting. Then, where the snowfield was firmer, he slid with hind legs trailing. The drift was softer lower down and he resumed wading, and where the slope became less steep, galloped in the snow as though he enjoyed the lark.

It also seems that grizzlies do not mind, and in fact apparently enjoy, a little sliding or "skiing," in this way resembling otters. Some slides are taken lying on one side. Slide marks on one slope showed that both a mother and cub had taken a rather long slide lying on their sides. On 17 May 1962, I watched a lone bear walk out on a snowfield, get mired, roll over to extricate himself, and then let himself slide while lying on his side. Near the base of the snowfield's slope he turned so as to face up the slope as he put on the brakes by digging in his claws. Coming to a stop, he continued to cross the snowfield on a lower contour line. His relaxed body and benign facial expression suggested he rather enjoyed the ride. Some snow trails consist of two parallel grooves coming, in some cases, directly down a slope. The bears making this kind of trail slide while standing in a skiing position. One bear appeared to have slid in this way for at least 200 feet (Fig. 12).

One spring I watched a bear descending a snowfield into a deep ravine and beginning to slide on his feet. Being a discreet bear, he felt he was sliding too fast and turned by digging in his forepaws until he was facing up the slope. He stopped by braking with all four feet. He made a couple of jumps to the brink of the steepest part of the slope, then, before sliding, he turned so as to face up hill and in this position slid into the ravine out of my view.

The following day I saw this same bear wading in deep snow that had drifted into willow brush on a gentle slope. Instead of continuing the wading, he lay on his side and rolled like a barrel over the snow and willow tops. After rolling over four or five times, he reached the edge of the willow patch and started walking, but after a few steps, he lay down again to progress by the rolling technique, rolling over four or five more times. I expect that the scratching effect of the rolling was about as much incentive as was the ease of progress it contributed.

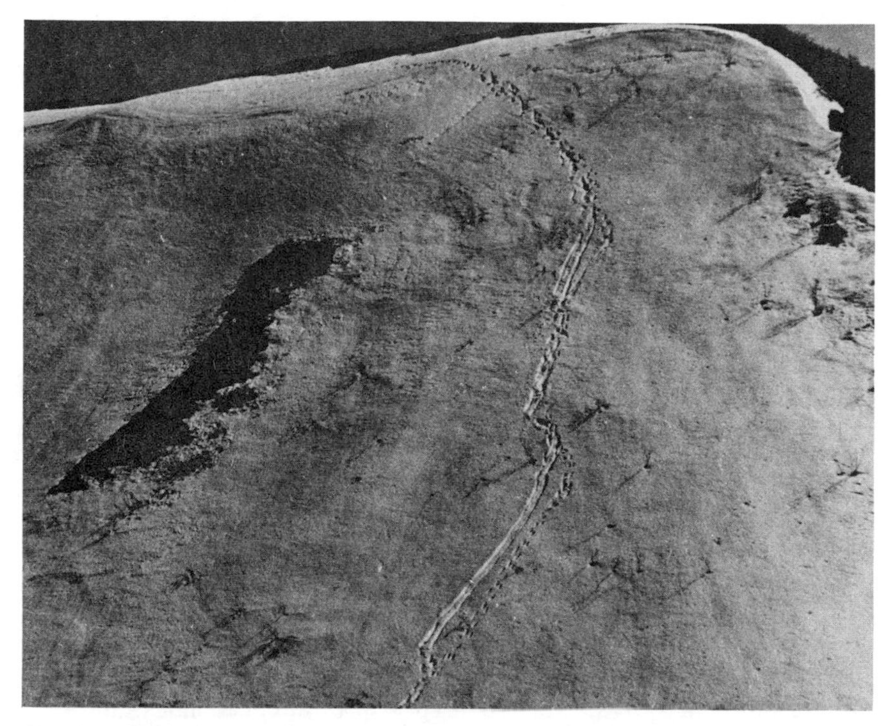

Fig. 12. Bear trail descending steep snow slope. One of the bears did some skiing.

To what extent snow and dirt bother the feet by adhering between the toes is not known. But one day in October, when a light skiff of snow had fallen, three balls of mud, each about 1½ inches in diameter, were picked up on a fresh trail. They had been pulled loose, apparently with the teeth, for a little hair was mixed with each one. Beyond the mud balls, a little blood was noted in the tracks. The mixture of snow and mud apparently had been just right for causing the mud to ball up.

The tracks of hind and front feet of the grizzly differ considerably. The main pad of the hindfoot is long; the claw marks extend only about 1 inch in front of the five toe-pad marks. The track of the forefoot consists of a short, broad pad mark posterior to five toe-pad marks. The tips of the long claws make marks about 2 inches in front of the five toe pads. If the track is deep, a small, rounded pad registers posterior to the main pad. Both front and hindfeet turn inward in walking.

Occasionally, the track of an individual may be distinctive. The main pad track of one hindfoot of a large male tapered toward the rear much more than it did in the other track. The difference was so obvious that this male's trail could be identified readily. A bear crippled on a front foot also left a distinctive track. The impression of the injured front foot was small and the marks of the unworn claws extended farther from the toes than in the normal foot.

On the river bars there are wet patches of mud or fine sand in which tracks show up well. However, a bear takes no pains to accommodate one with clear tracks. One day on the Toklat River bar, I saw a mother and yearling cross the river and move up along the bar. I followed to look for tracks, but even though there were many moist patches of mud scattered about, ideal for track impressions, they had all been avoided and not one track was seen. When one does find tracks either in mud or snow, they are often too ill-defined for accurate measurements. Sometimes a good front track may be found but no hindfoot track suitable for measurement. A track may be found that is satisfactory for measurement of width but not for length, so that one must examine several before finding measurable tracks. A series may vary in size because of slippage, depth of impression, rate of travel, or character of ground or snow. Hence, measurements obtained from a short trail may be rather miscellaneous in character and incomplete.

I have listed track measurements for both sexes and for cubs of different ages (Table 4). The claw marks are included in the measurements unless otherwise stated. In those for cubs there may be some variation due to differences in size between cubs of different families, and even in the same family. Tracks of the cubs in the different age groups tend, of course, to be larger in the fall than they are in the spring, due to the summer growth of cubs.

Table 4. Measurements of grizzly bear tracks. Single entries indicate measure of single tracks; others are a range from several tracks or, for cubs, from two litter mates.

	Front foot (inches)		Hind foot (inches)		Date
	Width	Length	Width	Length	
Adult female	5–5½		5	9	
Adult male	6–6¾		6–6½	10¾–12	
Spring cub	3	3⅝			25 July
	3				23 June
	3⅜–3¾	3¾ without claws		6¼–7	Aug.–Sept.
Yearling cub	4			7	10 June
	4			7	17 Aug.
	3¾–4			7⅛–8	16 Sept.
	3½				3 June
Two-year-old cub	4¼–4½				May
	4½–5		4½	9⅝	May
	4¼–4½	7¾	4¼	7–7½	3 June

Occasionally, one encounters a bear that is limping. He may have only a limp, or he may not use the foot in walking, or he may use it when walking but carry it when loping. Some bears have a permanent limp, while others may recover. In a later section dealing with grizzly–porcupine relationships, it is pointed out that a bear sometimes makes contact with a porcupine as indicated by quills sticking in the face or in a paw. A crippled bear that was killed had quills inside the crippled foot; apparently the quills had caused a permanent injury. A 2-year-old cub was stuck with quills, both in its face and one paw. Some other crippled bears will be described briefly in this section, for it would not be fair to the porcupine, or perhaps to the bear either, to list them all in the porcupine discussion.

On 13 June 1959, I saw a large, dark male with a severe limp of the left front foot. The elbow on the injured leg extended outward abnormally, and the foot turned inward excessively. The limp remained the same for the several weeks that the bear was seen. In 1963, this male grizzly was seen again, still with the same pronounced limp in the left front foot; obviously this bear had a permanent foot injury. One of two females mated to this male in 1959 limped on a hind leg. A round sore, the size of a dollar, could be seen just above the heel. When she galloped, the crippled foot was not used. She was first seen 17 June. On 1 July her foot had improved but she still carried it when she loped. On 4 July I saw her licking the sore spot; she limped a little but had improved, and by 10 July she seemed to have recovered. Over the years I observed several other bears, mainly cubs, with injured limbs. Of eight lame bears observed, six had injuries to a forefoot.

Injuries must be fairly rare and the causes of most is not known. Imprudent encounters with porcupines are certainly one source of such damage. Perhaps a cub occasionally is hurt slightly during over-exuberant play, especially with its mother, although I saw no evidence of this in several instances where females played roughly with their cubs. Sometimes a female, single-mindedly digging out a ground squirrel burrow on a rocky slope, will send large rocks flying below her, narrowly missing a cub. One spring cub was hit and rolled over by a rock about a foot in diameter in this situation, but did not seem to be injured.

Aside from their tracks, trails, and scats, "bear trees" are the other main sign of the presence of bears.

There is more than one kind of bear tree. When my brother and I were studying elk in northern Wyoming, we occasionally discovered trees with the bark torn loose near the base of the trunk. Species chiefly affected were the smooth-barked firs, but many lodgepole pines also were involved. The first time we encountered these trees we examined them closely and learned that the work was done by bears, apparently black bears. They were feeding on the cambium layer that carries the sap, seeking the raw syrup. The outer bark was bitten into and severed near the base of the trunk and the bark then pulled loose and stripped upward 3 to 4 feet or more so as to expose the cambium. Using the incisors, the cambium was scraped off, the close-set teeth leaving long, vertical, parallel grooves on the white, barkless trunk. The outer bark was pulled loose in several strips, occasionally around the entire trunk, so that when the bear was finished, these loose strips, attached 3 or 4 feet from the base, hung around the trunk like a grass skirt. Where such trees were found, there usually were several of them scattered in the area, sometimes 25 or 30. This suggested that once a bear tastes the "sweet" cambium delicacy he is reluctant to return to a more substantial diet. In some cases this feeding sign was both fresh and old. Perhaps the old sign was a reminder to a passing bear. In later years I also noted this feeding sign in the Olympic Mountains. We called them "bear trees."

When a black bear climbs an aspen tree, he leaves tracks. The claw marks heal over with scar tissue which remains for the life of the tree—a picturesque pattern registering a bygone event. These aspens are not at all uncommon in black bear–aspen country.

There is another kind of bear tree associated with both black and grizzly bears. These are trees situated so conveniently that they serve frequently as back scratchers. A lone tree along a trail or on the edge of a river bar is sure to be patronized often. Where many bears travel, even over a trail through a woods where numerous trees are available, there may be any number of bear trees showing signs of repeated use. In the Wood River country several miles east of McKinley National Park, I once followed for some distance a deeply worn bear trail through spruce woods bordering a high, perpendicular river bank. Bears passing up and down stream on that side of the river were somewhat hemmed in by the precipitous bank, sufficiently so that they generally used the trail. The traffic was so heavy that not only were the individual bear steps deeply worn, but much-used bear trees were closely spaced. I believe that the power of suggestion has given all bear travelers numerous itches, with the result that itching and rubbing on trees has increased through the years. I must add that all this bear sign on the crooked trail also increased my alertness, for at each turn I visualized a bear close enough for mutual embarrassment. Trails such as these, however, are rather scarce in the park, but along the Teklanika River there are short stretches of trails bordered by trees that show wear and have bear hairs embedded in pitch and lodged in the bark (Fig. 13).

A bear tree may show scratches and tooth marks 6 or 7 feet up the trunk, and limbs have been broken off at 7½ feet from the ground. In time the rubbing wears away patches of the bark. The ground is often worn smooth at the base of the tree. As the bear stands erect on hindfeet, with his back or stomach against the tree, or sits on haunches, he may

Fig. 13. A bear tree, used for rubbing.

bite randomly at the trunk, which, if the tree is slender, may produce two notches at different heights, roughly 3 and 5 feet. In a thorough scratching, the back, sides, rear, stomach, head, and neck are all massaged. Occasionally, the maneuvers suggest the latest "twist" dances as practiced by young people. One bear, standing on hind legs against a pole, raised and lowered herself, wriggling her body as part of the down movement to add to the effect.

In treeless country a large boulder is an excellent substitute for a tree. On Sable Pass I watched bears use a couple of poles that were lying on the ground, rolling on them so as to treat various parts of the anatomy. A few times I have seen a bear roll on the ground with much wriggling to get satisfaction, or to sit on haunches and rub his rear parts, which seemed to require excessive effort. Occasionally I have noted signs indicating that a bear had rubbed against a wooden bridge-railing and bitten off large slivers. Tall, stout willow brush is utilized sometimes. Frequently, bears rub on desirable sharp edges of log cabins and leave hairs. One morning, a large male rubbed against a log supporting the porch of Igloo Creek cabin and pushed it off its base. As they walk, bears often straddle brush or small spruce trees to scratch ventrally. No opportunities are overlooked.

Some of the literature suggests that trees are used by a bear to show other bears how high he can reach, hence how big he is, as a sort of warning to all to keep away from his domain. But in the first place, a grizzly does not lay claim to a domain. During the breeding season a bear tree might, I suppose, incidentally impart forcefully to a male the information that a desirable female passed that way. However, all observations indicate that the primary and conscious use of bear trees is for massaging.

When grizzlies encounter a pond in their travels, they may often wade in to lie down or take a swim, as though to cool off. They may also quench their thirst.

On 1 July 1940, what appeared to be a large male entered a small pond about 60 yards wide that lay in his line of travel. He splashed around a bit and swam across with body well submerged and nose pointed above the water. When he reached shore, he galloped for perhaps a half-mile.

On 8 September, 1939, I saw what appeared to be a young bear acting strangely. He hurried up a slope, then galloped down the ridge to Big Creek where he drank, then ran splashing down the middle of the stream. He was coming toward me, only 100 yards away, so I moved up the slope. When he came to my tracks, he reversed his direction, galloped up the stream, then over a ridge. He seemed to have enjoyed splashing his way down the stream, but the extreme exuberance was puzzling.

On 1 August, 1940, a female grizzly and spring cub walked along a small creek, feeding. It was a warm day and the female walked with open mouth and panted loudly. Three times she and the cub entered the creek, each time lying briefly in the water, obviously to cool off.

On 22 August 1950, Walter Weber and I saw a mother bear lying in a deep spot of Igloo Creek near Sable Pass, apparently cooling off.

On 15 September 1951, after a bear had rubbed his back on a pole, he sat down in a small puddle and appeared to wash his face with a paw.

On 24 July 1953, a mother and yearling were in a pond south of Cathedral Mountain. Part of the time the mother was completely submerged. They played a little in the water. The yearling soon went ashore but the mother remained in the cool water for several minutes.

On 28 May 1960, a large male entered a hollow and meandered around, attracted by the scent of meat. Earlier, two wolves were reported to have fed on something in this hollow and apparently had left only the scent of the food. After much frustration, the male walked into a pond in the hollow and lay with only his head and hump showing. A few times he submerged his head. Upon leaving the pond, he shook himself vigorously, did more searching for the source of the scent that was apparently strong in his nostrils, then re-entered the pond. When he left, he walked away without shaking, his wet hair flattened and dripping.

On 30 July 1962, the larger and more active of two yearlings entered a pond and swam and played for 3 or 4 minutes. The smaller yearling watched from the shore. He finally waded close to the shore, but only briefly. The mother continued to graze in the green hollow while the cubs were at the pond.

On 21 July 1953, a young bear spent several minutes wallowing in a pond, part of the time submerged. When he came out he seemed refreshed, approached a pole, bit into it, and then frisked away, galloping energetically. A bath often seemed to make the bears feel like romping.

On 27 August 1961, I watched two 2-year-old cubs that were in the process of separating from their mother straying off by themselves. After feeding for about 4 hours, they moved down to a small creek. One of them found a deep hole and waded in until the water covered all but his head. The other cub walked along in the middle of the stream for 30 yards and moved over to a green hollow to feed. It was a bright day, rather warm. I expect the cubs, in this rather casual manner, tarried in the water to cool off.

On 19 June 1955, a road worker watched a mother with spring cubs cross Igloo Creek, which still had some overflow ice protruding over the water in places. The spring cubs walked back and forth on shore, bawling, afraid to cross the rapid creek. One started across and was washed under the ice, but emerged farther downstream and managed to make the crossing. When my informant left, one of the cubs was still walking back and forth on shore, part of the time on his hind legs.

When streams are low, bears splash across without hesitation. But they recognize deep, fast water. On 14 June 1962, a lone bear, coming to a deep, rushing channel, stopped to ponder and walked along the edge circumspectly before entering. He was carried 50 yards downstream before reaching the other side.

The above notes are typical of the behavior often observed in the park. Along the coast of Alaska both black and brown bears spend much time in the water catching salmon, an activity not available to the McKinley National Park population of bears.

3
Range and Movement

Home Range

My data on the home range of grizzlies pertain chiefly to family groups, mothers and cubs, because they generally can be identified readily throughout the season and from one year to the next (Fig. 14).

In making identifications there are several helpful family characteristics. The mother may be blackish, chocolate, brown, or light tan (blond). She may have special markings such as a light or dark face and, in the case of blonds, some variation in the dark stripe between the shoulders and along the back. There may be one, two, or three cubs in the family. Moreover, the age of the cubs narrows the possibilities. They may be spring cubs, yearlings, 2-year-olds, and, occasionally, 3-year-olds. Spring cubs may be blackish or brownish and may or may not have

Fig. 14. Mother grizzly and yearling crossing a late spring snow patch.

a white stripe on one or both sides of the neck; the length and breadth of the white stripe, when present, may vary. The yearling and 2-year-olds also vary in color. When there are two or three cubs in a family, they may all be dark or all blond, or one may be much darker or lighter. The size of the individual cubs, when there are two or three in a family, may differ. The various combinations of these characteristics in a family group usually serve for ready identification. Occasionally, two families on the same range may be similar, especially when each family has only a single cub. Then greater familiarity with the family is necessary for identification.

Sometimes, special characteristics are helpful. For instance, one large male was missing an ear; another old male was permanently lame on a foreleg; a female limped on a hindfoot; two lone bears had severe limps; a few cubs limped, at least for a few weeks; a young bear had a scar below a hip. Many families were so well marked and seen so often that they became familiar to several people who were visiting the park for prolonged periods of time.

In some studies, grizzlies have been marked with ear tassels and had radio transmitters attached to them. Elk in Jackson Hole wear collars of various hues, moose are eartagged, and I have seen trumpeter swans wearing pink plastic collars. Many sensitive people who are sincerely interested in preserving wilderness are opposed to the use of such techniques in an area devoted to esthetics and spiritual values. The observation of tassels in the ears and the knowledge that the bears have been manhandled systematically destroy for many people the wilderness esthetics for an entire region. We might imagine a situation so critical that such intrusive, harmful techniques would be necessary. But in the case of the grizzly in McKinley National Park the added information obtained does not merit the sacrifice of the intangible values for which parks are cherished. In our wilderness parks, research techniques should be in harmony with the spirit of wilderness, even though efficiency and convenience may at times be diminished.

Because much of the country where bears were observed is treeless, frequent sightings were possible. However, they often were hidden in hollows and ravines or obscured by willow brush, but a little patience usually revealed them. A bear taking a nap could be hidden for an hour or two even though it was not far away.

I followed no daily routine in gathering home-range data; most observations were made incidentally in the course of general field work that often involved other species. More sightings of bear families could have been made if their ranges had been visited more frequently or if time had been devoted to looking for them.

The data were secured over a long period. The first notes on home range go back to 1922 and 1923, but most information was obtained after 1950.

A number of the home-range records pertain to the Sable Pass area because it is a favorite range for bears and visibility is good, but many observations were made elsewhere, especially on Igloo Mountain, the Polychrome Pass area, and the country westward to the Toklat River. Some ranges involved two or more of these areas.

A few families were observed over a period of 3 or 4 months but their total range from the time of their spring emergence from the den to their re-entry was not obtained. Bears were denning throughout the area where home-range observations were being made, so perhaps some of them were denning not far from where they were seen most often. One den, dug in July, was occupied later; this suggests that this bear denned in the middle of its summer range. No doubt there is great variation in the total extent of the ranges of bears and in the distance they cover from a denning site to the range they occupy for most of a season.

Many of the families were seen so often in a stretch of country 4 to 6 miles long by 2 or 3 miles wide that it was apparent that they wandered little farther during a period of several weeks. The ranges of all individuals noted were probably larger than indicated by my records. On one occasion a lone bear made a trek of over 20 miles in one day. One family was seen in an area over 18 miles in length. Many of the ranges were long and narrow because bears tend to remain in a valley, but some individuals were known to use two watersheds separated by a high ridge. Most of those seen on the eastern slopes of Igloo Mountain moved regularly over to the Big Creek drainage on the west side of the mountain and did much of their foraging there.

In the early spring, lone bears were observed traveling along ridges and apparently covered much ground. Some observations by William Nancarrow (pers. comm), however, suggest that families remained close to the den for a few weeks after emerging. A female and her two yearlings that he observed were seen daily within 100 to 200 yards of their den between 7 and 22 April.

The effect of seasonal food habits on home ranges varies. Some bears will remain in the same general area throughout the rooting, grazing, and berry-eating seasons. Over most of the park, the foods are sufficiently dispersed and intermingled to permit them to do this. Other bears may shift ranges with the food seasons. Thus the ranges of different bears may coincide for only one seasonal food period. There were families on Sable Pass that arrived at the beginning of the grazing period and moved elsewhere to feed on berries. Other families were in the area through most of the food seasons. Each family had its own movement pattern which, in some cases, also showed variation from year to year.

Weather may have some effect on the movements of bears. On a few occasions when they appeared on Sable Pass for the green forage and found none or very little because of the late season, they moved to lower, adjacent country and did not return. A failure of the berry crop also affected individual home ranges, causing bears to wander more widely.

Home-range data were gathered for a number of families. One mother was seen in 4 successive years with two sets of cubs. Two other mothers were seen followed by a single cub for 4 years. Eight families were seen over a 3-year period, that is, during the period they were followed by a set of cubs. Twenty-seven families were observed during 2 years. These data are shown in Table 5. Some data on home range were obtained from observations of 69 families seen two or more times during a single season.

The information gathered indicates that families tend to occupy the same areas from year to year. Except for the snow-covered, high mountains, the entire park is bear country so there are home ranges of various shapes and sizes over most of the area.

The data on the home ranges of several families and of a few other bears will be summarized to show the nature of the information secured. Locations referred to are noted on Figure 15. Milepost numbers refer to mileage from the McKinley Park railroad station.

Three Mothers Seen Over Four-Year Period

Female on Sable Pass: On 17 June 1959, I saw, for the first time on Sable Pass, a brown female with two yearlings. (In 1958 I was not in the park so had no opportunity to see the family when the young were spring cubs.) I saw the mother not only in 1959 but also the following three summers. She was recognized readily because of the deep brown color over most of her body. She behaved as though she was oblivious of humans. One of the yearlings was dark brown like the mother and a little larger than the other cub who was straw-colored or blond. The blond cub was a female and seemed to have a rather pointed muzzle. At first, I assumed that the brown cub was a male, but as it grew older it also appeared to be a female; however, I never ascertained its sex. The muzzle of the brown cub was a little heavy; the facial line was rather straight, creating a profile sufficiently distinctive to cause a friend to refer to the cub as ''profile.'' The blond cub was always the more active and had a quick step; she always strayed farther from the mother and moved about a great deal as she fed. The brown yearling seemed somewhat phlegmatic and a follower. During the four summers that I watched the cubs, these individual behavior traits prevailed. (I observed the cubs three summers after they separated from the mother).

In 1959, the family was seen on Sable Pass on 31 days between 17 June and 4 August. It ranged over an area, so far as my observations indicate, 7 or 8 miles in length and 2 miles wide. On some days the family moved less than one-half mile, but on one occasion I saw them make a 2-mile trek without stopping, moving away from the vicinity of a large male grizzly. This family came to Sable Pass at the start of the grass and herb season and departed early in the berry season. (Some families found berries and roots in the Sable Pass area.) I never determined where this mother fed on roots in the spring or where she spent the berry season, but one year I saw her, a few weeks after she had left Sable Pass, about 5 miles to the north near lower Igloo Creek. In 1959, the family left Sable Pass on 4 August, walked northward on a high contour paralleling Igloo Creek for about 2 miles, and was last seen going over a ridge toward Big Creek. The family had moved down into lower country for the berry season, and was not seen again until the following year.

In 1960, I first saw the family on 18 May, a month earlier than the previous year, as they emerged from a canyon of Cathedral Mountain. I discovered the family on the move at Milepost 35½. During the day, they crossed Igloo Creek and fed, then continued on to Milepost 41, taking a shortcut over two sizeable ridges.

On 20 May, 2 days later, the two 2-year-old cubs were seen about 3 miles farther west on the west side of East Fork River. The mother had deserted them apparently to consort with a male. (*See* Mother–Cub Separation.) The movements of the cubs in 1960 and the following 2 years will be discussed separately, but it may be stated here that they remained in the Sable Pass–East Fork area. The mother was seen again on 27 May, when she apparently had finished breeding. She was seen rather infrequently during the summer, which suggests that she wandered more widely when alone than she had the previous summer with the cubs. During the summer, she was seen on 18, 27 May, 22, 27, 28 June, 29, 30 July, 22 August, and 20 September. In 1960 she came to Sable Pass a month or more earlier than in other years and departed about 6 weeks later. With the exception of the 18 May sighting, she was seen always on Sable Pass over an area about 4 miles long and a mile across.

On 13 July, 1961, this female appeared on Sable Pass with two spring cubs. Between then and 29 July she was seen nine times on the pass. With the advent of the berry season, she moved a mile down Igloo Creek where she and her cubs were seen feeding on berries 6 days between 6 and 15 August. She then disappeared, apparently moving down country to the north as she had done in 1959. Her total range during the period she was observed was about 7 miles long and at least 2 miles wide.

The female and her two yearlings were first seen on Sable Pass on 24 June in 1962. She was observed 15 times between 24 June and 31 July at rather regular intervals in a narrow strip 4 miles long. On 31 July she

Table 5. Summary of home-range data for grizzly bear families seen for two or more summers.

Family	No. of days seen	Period	Size of range (length in miles)	Locality
two yearlings	31	17 June–4 Aug. 1959	7–8	Sable Pass–Igloo Cr.
two 2-year olds	9	18 May–20 Sept. 1960	6	Sable Pass–Igloo Cr.
two spring cubs	15	13 July–15 Aug. 1961	7	Sable Pass–Igloo Cr.
two yearlings	16	24 June–25 Aug. 1962	12	Sable Pass–Igloo Cr.
one spring cub	23	25 May–22 Sept. 1963	9	East Fork–Toklat
one yearling	22	28 May–20 Sept. 1964	9	East Fork–Toklat
one 2-year old	33	1 June–6 Sept. 1965	18	East Fork–Toklat
one 3-year old	3	30 May–4 June 1966	2–6	Toklat
one spring cub	4	29 June–1 Sept. 1964	5–6	Cathedral Mt.–East Fork
one yearling	15	17 May–22 Aug. 1965	10	East Fork–Toklat Riv.
one 2-year old	9	11 June–4 July 1966	3	Toklat River
one 3-year old	3	10 June–16 June 1967	4	Toklat River
three spring cubs	—	reliably reported 1938		Polychrome (48 Mile)
three yearlings	4	23 May–30 Sept. 1939		Polychrome (48 Mile)
three 2-year olds	16	4 May–23 Sept. 1940	8	Polychrome (48 Mile)
one spring cub	5	20 June–7 Aug. 1953	5–6	Igloo Cr.–Sable Pass
		no field work 1954		
one 2-year old	9	30 May–22 July 1955	5–6	Igloo Cr.–Sable Pass
two spring cubs	32	23 May–22 Sept. 1960	5–6	Sable Pass
		not seen 1961		
two 2-year olds	1	28 May 1962		Tattler Cr.
two spring cubs	19	17 July–17 Sept. 1961	4	Sable Pass
two yearlings	27	2 June–5 Sept. 1962	6	Sable Pass–East Fork
two 2-year olds	3	15 June–17 June 1963	1	East Fork

two spring cubs	18	4 June–28 Sept. 1961	3	Igloo Mt.–Tattler Cr.
two yearlings	28	12 May–17 Sept. 1962	7	Cathedral Mt.–Sable Pass
two 2-year olds	19	24 May–2 Sept. 1963	6	Cathedral Mt.–Sable Pass
one yearling	17	2 June–21 Sept. 1961	2	Cathedral Mt.–Igloo Mt.
one 2-year old	20	31 May–11 Sept. 1962	2	Igloo Mt.
one 3-year old	2	30 May–31 May 1963		Cathedral Mt.
two spring cubs	8	3 July–12 Sept. 1962	5–6	Cathedral Mt. East Fork
two yearlings	3	27 Aug.–25 Sept. 1963	5	Cathedral Mt. East Fork
two 2-year olds	2	30 May–31 May 1964		Igloo Cr.
two spring cubs	20	9 June–22 Sept. 1964	3	Sable Pass
two yearlings	12	29 May–30 Aug. 1965	3	Sable Pass
two 2-year olds	5	4 June–9 June 1966	1	Tattler Cr.
two spring cubs	Several	July–19 Sept. 1922	3	Head of Savage Riv.
two yearlings	Several	15, 17 etc. July 1923	3	Head of Savage Riv.
		no field work, 1924		
three spring cubs	1	1 Aug. 1940	4	Polychrome Pass
three yearlings	1	8 July 1941	4	East Fork Riv.
		no field work 1942		
one spring cub	19	5 June–9 Oct. 1940	13	Igloo Mt.–Polychrome
one yearling	3	25 May–1 June 1941	1	East Fork River
		no field work 1942		
three spring cubs	1	8 Aug. 1940		4 miles down Toklat Riv.
three yearlings	1	23 June 1941		4 miles down Toklat Riv.
		no field work 1942		
two yearlings	1	17 May 1947		Polychrome Pass
two 2-year olds	2	14, 17, June 1948 (edge of range)	1	Polychrome Pass
one spring cub	4	3–7 July 1948	1	Sable Pass
one yearling	1	24 July 1949		Sable Pass

Table 5. (continued)

Family	No. of days seen	Period	Size of range (length in miles)	Locality
two yearlings	3	4 Aug.–14 Oct. 1948	1	Igloo Mt.–Cathedral
two 2-year olds	1	14 May 1949		Igloo Mt.
two yearlings	1	11 June 1948		Teklanika Riv.
two 2-year olds	1	11 June 1949		Teklanika Riv.
two spring cubs	17	26 June–23 Sept. 1950	7	Sable Pass
two yearlings	9	28 June–15 Sept. 1951	6	Sable Pass
one yearling	7	9–28 July 1955	1	Sable Pass
one 2-year old	17	22 May–23 July 1956	5	Sable Pass
two spring cubs	9	2 June–13 July 1955	4	Igloo Creek
two yearlings	7	18–27 May 1956	2	Cathedral Mt.
		no field work 1957		
one yearling	2	10–19 June 1955	1	Sable Pass
one 2-year old	3	9–27 June 1956	2	Sable Pass
three spring cubs	1	14 July 1955		near Toklat Riv.
three yearlings	3	12–29 June 1956	4	Polychrome–Toklat
		no field work 1957		
two yearlings	3	9 Aug.–1 Sept. 1959	2	Cathedral Mt.-Igloo Mt.
two 2-year olds	1	23 May 1960		Cathedral Mt.
		no field work 1958		
two yearlings	13	9 July–10 Aug. 1959	5	Sable Pass
two 2-year olds	12	30 June–11 Aug. 1960	8	Sable Pass–Polychrome
		no field work 1958		
two yearlings	16	25 May–9 Aug. 1959	6	Sable Pass
two 2-year olds	1	23 May 1960		Sable Pass

two spring cubs	4	28 July–10 Aug. 1959	3	Toklat River
two yearlings	2	26 July–25 Aug. 1960	1	near Toklat Riv.
two spring cubs	9	30 May–12 June 1960	2	Polychrome Pass
two yearlings	2	13–16 May 1961	2	Polychrome Pass
one spring cub	6	11 July–15 Sept. 1960	2	Highway Pass
one yearling	3	6 June–14 Sept. 1961	1	Highway Pass
two spring cubs	4	24 June–2 Sept. 1960	2	Highway Pass
two yearlings	2	29 May–28 July 1961	1	Highway Pass
one yearling	21	19 May–21 Sept. 1961	7	Sable Pass
one 2-year old	9	19 May–7 Sept. 1962	8	Sable Pass
one spring cub	1	29 July 1963		Thorofare Riv.
one yearling	2	13, 23 July 1964	2	Thorofare Riv.
two yearlings	2	17–21 Sept. 1963	½	near Teklanika Canyon
two 2-year olds	3	24–30 May 1964	1	near Teklanika Canyon
one spring cub	23	25 May–22 Sept. 1963	10	East F. River–Toklat Riv.
one yearling	22	28 May–20 Sept. 1964	9	Polychrome–Toklat Riv.
two yearlings	7	17 June–31 July 1966	5	Toklat Riv.
two 2-year olds	8	9 June–16 July 1967	3	Toklat Riv.
one spring cub	5	2 Aug–31 Aug. 1969	5	Igloo Cr.–Sable Pass
one yearling	3	15 June–23 June 1970	5	Igloo Cr.–Sable Pass
one 2-year old	16	2 June–15 Aug. 1969	6	Igloo Cr.–Sable Pass
one 3-year old	1	5 June 1970		Igloo Cr.
		no field work after 4 July 1970		

Fig. 15. Map of the study area, Mt. McKinley National Park.

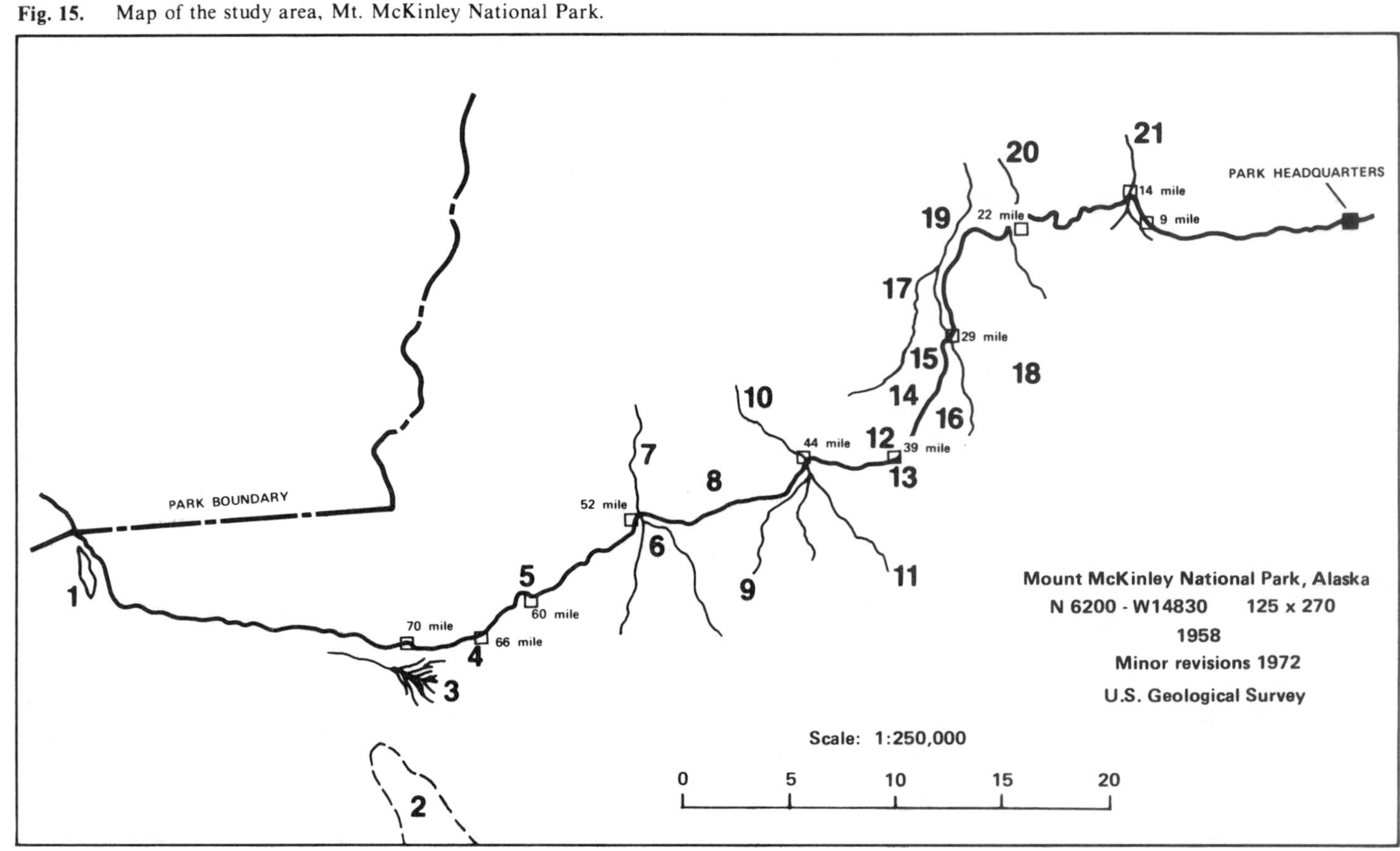

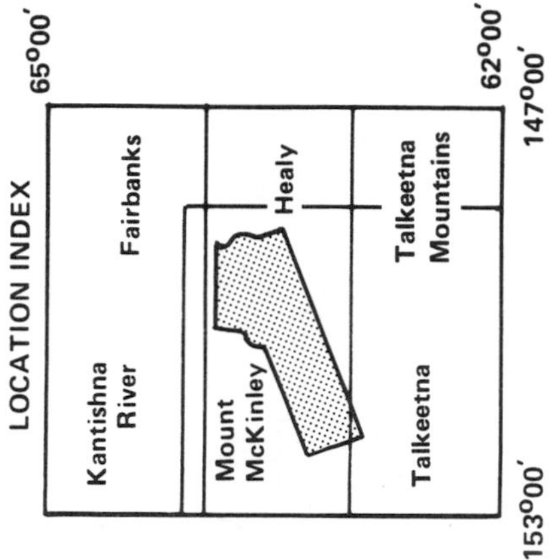
LOCATION INDEX
65°00′
62°00′
153°00′
147°00′
Kantishna River
Fairbanks
Mount McKinley
Healy
Talkeetna
Talkeetna Mountains

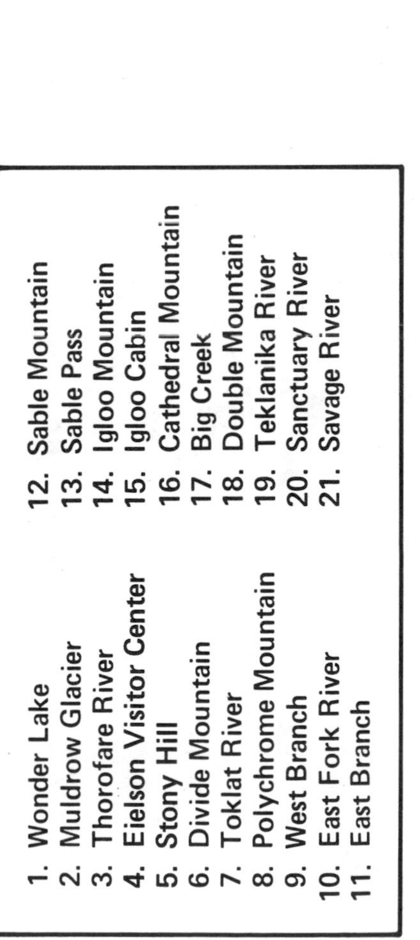
Place Names - Mt. McKinley
1. Wonder Lake
2. Muldrow Glacier
3. Thorofare River
4. Eielson Visitor Center
5. Stony Hill
6. Divide Mountain
7. Toklat River
8. Polychrome Mountain
9. West Branch
10. East Fork River
11. East Branch
12. Sable Mountain
13. Sable Pass
14. Igloo Mountain
15. Igloo Cabin
16. Cathedral Mountain
17. Big Creek
18. Double Mountain
19. Teklanika River
20. Sanctuary River
21. Savage River

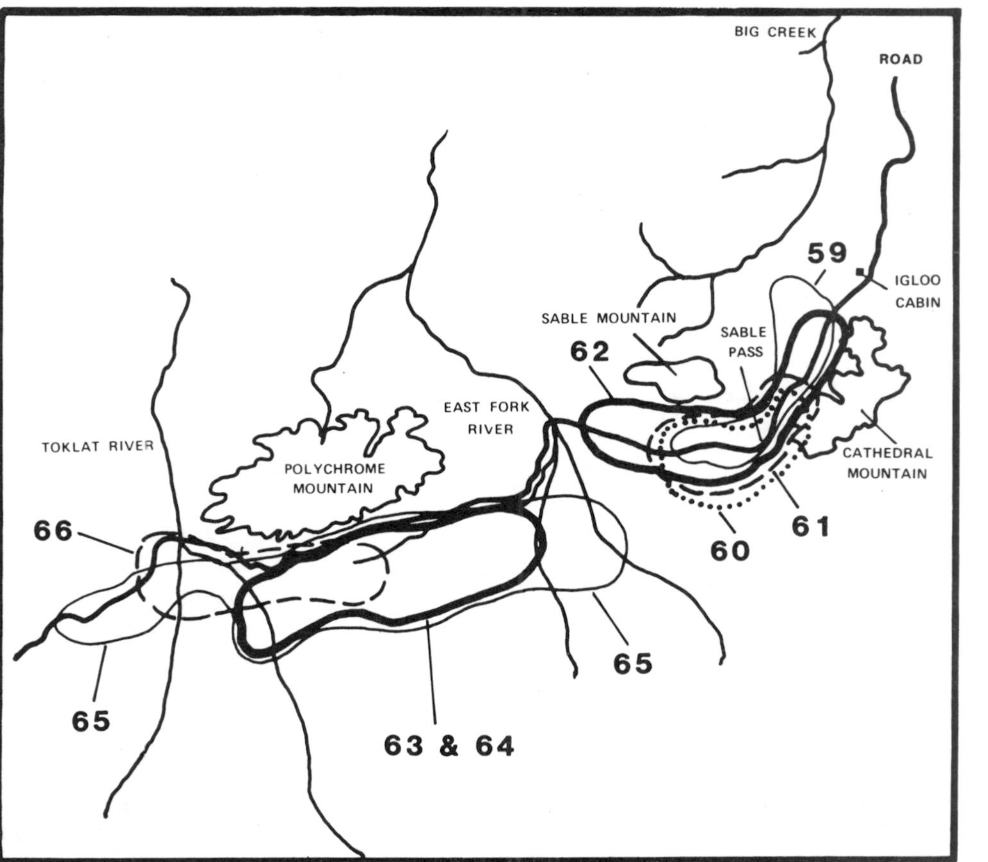

Fig. 16. Approximate home ranges for two females recognized for periods of 4 years. Number 1 is a female seen initially with two yearlings in 1959 (see page 36). Number 2 is a female with a single spring cub in 1963 (see page 45).

moved down Igloo Creek 2 miles. As in 2 of the 3 previous years, she left Sable Pass near the beginning of the berry season. On 25 August I saw her about 5 miles north of Sable Pass, at a caribou carcass. She was observed in a stretch of country about 12 miles long. In 1963 I did not see the family (Fig. 16).

Thus we had a female returning to Sable Pass for 4 successive years. She seemed to have a rather definite pattern in her movements. Three of the 4 years she foraged in the pass only during the midsummer season. Other bears came early and stayed later, just as she did in 1960. It was not known where she denned, so the size of her total range was not known.

Mother and Cub Four Years on Same Range: In 1963, a dark-brown mother and a spring cub often were seen ranging from East Fork River to Toklat River, in an area about 9 miles long and 3 or 4 miles wide. I saw the family 23 times between 25 May and 22 September, and other observers were constantly reporting them. Generally, they were seen in a stretch of country about 5 miles long and a mile or two wide. On 23 July I watched the family make a 3-mile trek, traveling steadily except for brief stops for ground squirrels or a few bites of green food. It appeared that the female had decided to leave the area because she disappeared northward, but she returned later to her usual range. On 7 September I saw the family traveling steadily up the east branch of the Toklat River on a high contour of Divide Mountain as though it was going places. Perhaps the mother was looking for berries which were scarce that year. After going about 2 miles, she crossed the broad river bar and started back on the opposite side of the river. The bears made a long gallop to get away from the vicinity of a lone bear and then settled down to a rapid walk. The cub lagged 300 yards behind and did not catch up until the mother lay down and waited for it.

I saw the family 22 times on the same range at rather regular intervals between 28 May and 29 September 1964, the last day I visited its range.

The family was first seen on 1 June 1965. On 3 June it moved 3 miles farther east than I had ever seen it. Hunting and chasing calf caribou was the cause of this extension of range. The following day it was back in its usual haunts. The family was seen 33 times between 1 June and 6 September. On the latter date it moved 6 miles west of the most westward point I had seen it, moving steadily and held up by only a few stops to excavate ground-squirrel holes. For the last mile that it was in view, it traveled steadily and disappeared north of Slide Lake. For most of the summer the family covered the same range as the previous 2 years, but the 3 June eastward movement and the 6 September westward trek increased the known extent of its range to about 18 miles.

I discovered the family digging roots along Toklat River on 30 May 1966. It also was seen there on 3 and 4 June. On 4 June photographers driving unhurriedly in a car to where the bears were going under a bridge gave them a real scare, causing them to hurry northward. I did not see them again but twice there were reports of a family 4 or 5 miles east of Toklat River, which may have been these same bears (Fig. 17).

This family remained on the same range through much of the rooting, grazing, and berry seasons. In 1965 the mother probably left the range earlier than usual because of the scarcity of berries.

A Family Shifts Range: A blond mother and her spring cub were seen four times from 29 June to 1 September 1964, between the south tip of Cathedral Mountain and the upper reaches of East Fork River, a distance of 5 or 6 miles.

The family was seen 15 times from 17 May to 19 July 1965, between East Fork River and Toklat River. The west boundary of the 1964 range had become the east boundary of the 1965 range. On 22 August I watched the family travel 5 miles westward from Toklat River, a move taking it beyond the summer range. The family was not seen again in 1965. Apparently, the bears were seeking berries which were scarce that year.

The family was seen nine times between 11 June and 4 July, 1966 along Toklat River within the area occupied during most of 1965. During 1965, 1966, and 1967, the family occupied a range adjacent to its 1964 range. The total range occupied in the 4-year period was a minimum of about 22 miles in length.

In June 1967 this female and her 3-year-old cub were sighted three times in the area where they were seen in 1965 and 1966. The female was seen breeding with a large male on 10 June, and he subsequently followed the cub, apparently a female, up the river and out of sight. Six days later, the mother and 3-year-old were seen together for the last time (Fig. 18).

Eight Families Seen Over Three-Year Period

Mother with Triplets on Range for Prolonged Period: In 1939 a female and three yearlings ranged in the Polychrome Pass Area throughout the summer. I saw them on only four occasions but they were seen frequently by members of a road construction crew camped at Milepost 48. The family was visiting the nearby camp garbage dump. I saw the bears a month before camp was set up and a month after it was abandoned (23 May to 30 September). The camp foreman, a reliable observer, reported that the family had been seen frequently in the same area in 1938 when the mother was followed by three spring cubs. The family was well known. In 1940 when the cubs were 2-year-olds, I observed the family in the same general area on 16 occasions from 4 May to 23 September. During these 3 years (1938–40), most of their time out of the den was spent in an area about 8 × 2 miles. Usually, their travels were much more circumscribed. Their wanderings throughout the various food pe-

Fig. 17. The dark brown female and her cub, now 3 years old and almost as big as the mother, received a scare from photographers and galloped off up the Toklat River.

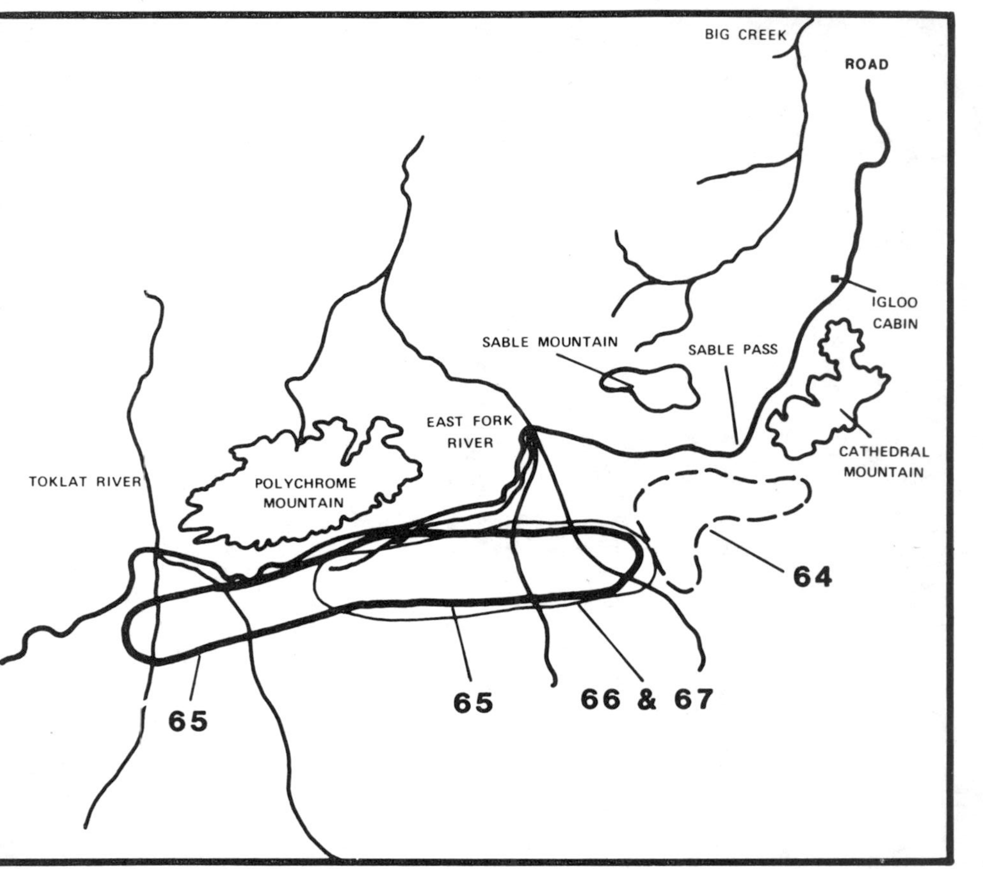

Fig. 18. Approximate observed home range of a blond female with one spring cub in 1964 seen the following 3 years (see page 46).

riods were confined to the same general area. The seasonal foods were intermingled in this range. When the garbage dump was in use, the movements of the bears undoubtedly were affected by it, and they may have been indirectly affected when it was not in use. But their travels and size of range were similar to families not affected by garbage. The three cubs apparently separated from the mother in autumn 1940. They were not seen the following year when they were 3 years old, and the mother, if seen, was not recognized.

Dark Mother and Single Cub: A dark mother with a spring cub were observed five times along Igloo Creek and on Sable Pass from 20 June to 7 August 1953. When first seen, on 20 June, they were traveling steadily up Igloo Creek, apparently on their way to Sable Pass from farther north. I was not in the park in 1954 when the cub was a yearling. Between 30 May and 22 July 1955 I saw the mother with her single cub, now 2 years old, nine times in the same general area. I left the park a week after my last sighting of this family. The family ranged over an area 5 or 6 miles in length during both years.

Dark Mother Shifts Range: A dark mother with two blackish spring cubs were seen 32 times from 23 May to 22 September 1960 in the Sable Pass area. She resembled a female that had mated in 1959 on the pass. Her known range was 5 or 6 miles in diameter. The frequent sightings by me and by others indicated that she did not wander much beyond this area. In 1961 the family was not seen. On 28 May 1962 I saw a dark mother with two dark, 2-year-old cubs that appeared to be this family, moving up Tattler Creek (near Sable Pass) toward Big Creek. Apparently, there had been a shift in the range after 1959, probably into the adjacent Big Creek drainage.

Blond Mother on Sable Pass for Extended Periods: A large blond female with two spring cubs were first noted on 17 July, 1961 on Sable Pass. Between 17 July and 17 September, the family was seen on this pass 19 times at rather regular intervals in an area about 4 miles across.

I saw the family digging roots sporadically from 2 to 21 June 1962 on the East Fork River bar along the western edge of Sable Pass. During the grass-eating period, mid-June to about the end of July, the bears were seen on adjacent Sable Pass, and during the berry season they moved over the same area and often were down on the river bar feeding on buffaloberry. Between 2 June and 5 September they were seen 27 times in an area about 6 miles long and 2 miles wide.

The family was seen on the bars of East Fork River digging roots on 15 to 17 June, 1963. This is where it had been seen in early summer 1962. The family was not seen again during the summer. It had spent two full summers on Sable Pass and had put in a 3-day appearance the third summer (so far as I could observe). It is possible that the female left to breed, thus altering home-range habits for the year. But they may also have shifted their range slightly, enough to keep them hidden from me.

Family Ranging between Igloo Mt. and Sable Pass: A blond female with 2 spring cubs was seen 18 times at regular intervals, between 4 June and 28 September, 1961 from Igloo Mountain to the base of Sable Pass (Tattler Creek) within an area about 3 miles in length. How much farther the family traveled or in what direction was not determined, but it was obviously moving about in a circumscribed area during the summer and fall.

This family was seen on 12, 14, and 15 May 1962 on Cathedral Mountain, across from Igloo Mountain. Between 12 May and 17 September the family was seen on 28 days. It ranged from the north end of Cathedral Mountain over the top of Sable Pass, in an area about 7 miles long and a maximum of about 2 miles wide. In 1962 the range had been extended to include part of the Sable Pass area.

In 1963 the family was first seen on 24 May near the north end of Cathedral Mountain where it had first been seen the year before. Both years the family fed on roots in this area. Later it moved to Sable Pass. During the period from 24 May to 2 September I saw the family at fairly regular intervals on 19 occasions, and it was reported on a few additional days by other observers. During these 3 years (1961–63) this family was known to range over the same general area for 3 or 4 months, and very likely was present before and after the periods reported here.

Cub Still with Mother When Three Years Old: I saw a blond mother with a very blond yearling on 17 occasions from 2 June to 21 September 1961. They were seen first in the spring near the north end of Cathedral Mountain where they spent a few days digging roots. The rest of the summer they generally were seen on Igloo Mountain and spent most of their time on the Big Creek side. In the fall they were last seen on 21 September near the north end of Cathedral Mountain moving toward Teklanika River. All observations were made in an area about 2 miles in diameter. The extent of their movements in the Big Creek watershed was not learned.

Between 31 May and 11 September 1962 the family was seen 20 times on Igloo Mountain. The movements were similar to those of the preceding year. Several times they were observed hurrying to the Big Creek side of Igloo Mountain. For example, on 11 August the mother and her 2-year-old were feeding on buffaloberry on the southeastern slope of Igloo. Later, the cub climbed a short distance and lay resting on a patch of grass. When the mother approached, he galloped ahead up the slope, continuing over one side ridge after another, sometimes returning to the top of a ridge to see if the lagging mother was following. She kept coming at a fast walk and occasionally broke into a lope. Sometimes she stopped to feed briefly on berries. Thus they traveled for 1 hour and 10 minutes before going over the top into Big Creek drainage. Three days later there was practically a repeat performance. The far side of the mountain seemed to be home to the cub.

On 30 and 31 May 1963, when the cub was 3 years old, the family was seen near the north end of Cathedral Mountain. Later the cub was seen alone. It is likely that the mother had deserted the cub so she could breed.

Blond Female with Two Darker Cubs: Between 3 July and 12 September 1962 a blond female with two dark, spring cubs were seen eight times. They ranged from the south end of Cathedral Mountain to the head of East Fork River, a distance of 5 or 6 miles. The family was seen three times between 27 August and 25 September 1963, from East Fork River to the south end of Cathedral Mountain, a distance of about 5 miles, and on 30 and 31 May 1964 the family was observed along Igloo Creek. Total range, according to my observations, was about 8 miles. This family apparently ranged chiefly toward the heads of Igloo Creek and East Fork River, where it usually would be out of view.

Late Spring Affecting Range of One Family: A dark mother with two spring cubs were seen 20 times in the Sable Pass area, from 9 June to 22 September 1964 (last day I was in the field). Between 29 May and 30 August 1965 the family was seen in the same area 12 times. Both years, according to my observations, the maximum extent of the range was about 5 miles. Most sightings were in an area 2 miles in diameter. This family was recognized easily because of the unusually wide, white collar of one of the cubs. The family was seen at Tattler Creek, within the area occupied the previous summer, on 5 days between 4 and 9 June 1966. Available grazing in the Sable Pass area in 1966 was unusually late because of the deep winter snow. This apparently discouraged bears and caused them to seek forage elsewhere that year.

Families Seen in Two Successive Years or in One Year Only

Home-range data were gathered on 27 families for two of the usual three summers that the cubs are with the mother. For 15 of the families it was not possible to get information for all 3 years because of my absence from the field in the year that the cubs were either spring cubs, yearlings, or 2-year-olds. In five cases the families were seen only when they were at one edge of their home range, which was chiefly beyond my usual travels; others were recorded in areas seldom visited or where the country was wooded and broken. Consequently, sparse data were to be expected for these families. In addition, my records indicate that 164 families were seen during only a single year. Of these, 69 were seen two or more times, but many of these observations were too fragmentary to warrant consideration here.

Home-range data for a few of the 27 families seen in 2 successive years, along with a few of the 69 families seen two or more times in a single year, will be summarized briefly.

Families Seen at Edge of Home Range: In some areas I saw some bear families that obviously were on the edge of their home range. Consequently, these data give little information on the extent of their wanderings but supplement the data showing that grizzlies have definite home ranges.

On 18 June 1939 I spent a memorable day watching wolves and caribou (the latter in migration) from a strategic point on a slope of Cathedral Mountain. I had a superb view of the fork in the Teklanika Valley. I discovered a mother with three spring cubs on the far side of the river and watched them during the day as they foraged. On the following day the family was seen again, this time at close range, for we met on the brow of a rise. The mother was so close I could see the patient expression on her face, as though she were waiting for the traffic to turn aside, which it did. She had moved a little over a mile from where she had been seen last on the previous day. On 20 June I did not see the family and after that date I was seldom in the area. On 8 August the family was discovered digging for a ground squirrel on the west slope of Cathedral Mountain, about 2 miles from where I had seen it in June. The squirrel captured, the family moved back toward the Teklanika River which apparently was the center of its range.

Another family that was seen several times on the west slope of Cathedral Mountain also appeared to be on the west edge of its range. The mother and two spring cubs were seen 10 times between 2 June and 13 July 1955 along a 3-mile stretch on the west side of Cathedral Mountain. When last seen, they were headed for the Teklanika River valley on the east side of the mountain. These bears were observed digging roots on the west slope of Cathedral Mountain seven times between 18 and 27 May 1956. On 27 May they moved around the north end of Cathedral Mountain toward the Teklanika River where they had gone the previous year, and were not seen again. (I was absent from the park in 1957.)

Several families seen on the southeastern slope of Igloo Mountain appeared to spend most of their time in the Big Creek drainage on the west side of the mountain.

On 22 June 1956 a blond mother and her blond yearling fed about 200 yards above Igloo Creek. When the mother became aware of me, she started up the slope of Igloo Mountain, the yearling moving out ahead, leading the way. High on the slope in the shale they passed close to mountain sheep that had moved to one side and stood watching. The bears paid them little attention, concentrated on leaving the country, and disappeared over the skyline headed for the Big Creek side. On 5 July the family was foraging high on the slope and, on seeing me, again hurried over the high ridge. This family was seen nine times in this area between 31 May and 23 August.

Also in 1956, another family, a rather blondish female with a dark yearling that ranged chiefly in Big Creek, occasionally was seen on Igloo Mountain. The family was seen six times between 15 June and 7 September. Both of these families were missed in 1955 when the young were spring cubs, and I had no opportunity to see the 2-year-olds because of my absence from the park in 1957.

On 9 August and 2 and 4 September 1964, a mother and two spring cubs were seen on Igloo Mountain. Their range was chiefly in Big Creek but they visited the east slopes of Igloo Mountain in search of berries. Some bears came over from Big Creek quite often, some seldom, and some perhaps not at all. The southeast slope was the edge of the range for most bears seen there, but for some this slope fitted into a different home-range pattern, and was the northern edge of a range that extended up Igloo Creek to Sable Pass.

Another section of the park where families obviously were seen at one edge of their range was the Polychrome Pass area. Here, the home-range pattern of some families was such that its main part was to the north of the road where they were soon hidden by the broken topography. From 9 to 21 June 1962, a mother and three spring cubs were seen three times north of the road on south-facing slopes, seeking the early green grass and herbs. In July they were reported on a few occasions a mile or two farther north. Between 12 and 27 August the family was seen four times on the flats to the south of the road where they had come to feed on buffaloberry. The blueberry and crowberry crops were so poor that bears were wandering widely in search of berries. Other families also put in an appearance on the flat to feed on buffaloberry. A mother and two spring cubs were seen there on 22 and 27 August, and a mother and a yearling, on 21 and 22 August. They apparently had come from the country to the north. These families were seen only during this one year.

A dark female and two blackish spring cubs were seen in the Polychrome Pass area nine times between 30 May and 12 June, 1960, in an area about one-half mile across. They were grazing the new growth of grass and herbs. On 12 August I saw them about 2 miles to the east. They were wary and hurried northward over a ridge. This same family was on Polychrome Pass on 13 and 16 May, 1961, in an area about 2 miles across. They were still ranging to the north. The family was not seen in 1962 which is not surprising because my observations the previous 2 years obviously were made on the edge of their range.

Families Traveling from One Seasonal Range to Another: A female and three yearlings were seen on a flat on the east side of Thorofare River on 31 May 1959. The three cubs frolicked and galloped at times, a little ahead of the mother. They were moving southward. On 30 August this family again was discovered at the base of Mount Eielson, about 2 miles from where they had been seen in May. Three or 4 inches of snow lay on the ground. The mother was digging ground squirrels, her cubs huddled about 100 yards from her, hidden by a growth of willows. Once she stopped digging, looked around, and dashed toward the cubs. She sniffed them as though to reassure herself of their identity and then returned to her digging. A few minutes later the cubs began to gallop westward across the high bench along the base of the slope. When a quarter mile from the female, she suddenly noticed them going away and followed at a lope. They all galloped for almost a mile, the cubs frolicking and the mother hurrying to overtake them. She caught up to one of them and later they all came together at a prospector's abandoned cabin which the two leading cubs had stopped to investigate. Again the cubs galloped forward and left the mother far behind, digging. Later, she galloped after them until I lost sight of them in the rough country over toward Muldrow Glacier. The cubs had much to do with the course of travel. My records may have been inadequate, but it appeared that both in spring and in autumn I had seen these bears as they were shifting from one seasonal range to another. On 16 September 1960, three bears, all the same size, were reported about 2 miles from where the three yearlings had last been seen in 1959. On 19 September I saw what seemed to be the same bears about 2 miles farther west from where they had been seen on 16 September. They appeared to be 2-year-olds and were possibly the three yearlings seen in 1959.

On 24 June, 1960 a mother with two spring cubs were seen at Highway Pass, traveling eastward toward the Toklat River. They seemed to be on their way toward the head of the Toklat River. On 30 August and 1 September, the family was feeding a mile east of Highway Pass and on 2 September it had moved westward over the pass. The family had been seen passing eastward in June and back westward in the fall. On 29 May 1961, the family again was seen on the east side of Highway Pass and was not seen again until 28 July when it showed up on the same pass. It appeared that this family was only observed in transit from the denning area to midsummer range, and again on its return.

On 4 and 11 August 1948 a female and two yearlings were seen on Igloo Mountain (southeast slope). Their range apparently centered on Big Creek on the west side of the mountain. On 14 October the family was seen moving to the east, away from the mountain. It apparently was leaving its summer–autumn range and going toward its denning site. In 1949 the family returned early to its summer range. The mother and her 2-year-olds were seen on 14 May feeding on roots and berries on the southeast slope of Igloo Mountain as they crossed snowfields on their way to the Big Creek side. I had failed to see the mother when she was followed by her spring cubs. The south slope of Igloo Mountain seemed to be one edge of her summer range.

Family Shifting Range from One Year to Another: A few observations indicated that a family had shifted its home range at least a few miles. Because there is so much joint occupation and overlapping of ranges, one would expect much more shifting of ranges than was observed. Yet there apparently is a strong tendency for bears to use the same ranges year after year, the ones with which they are thoroughly familiar, but there obviously are minor variations through the years, and a shift to adjacent terrain probably is not rare.

A rather dark female with a dark yearling cub were seen 21 times from 19 May to 21 September 1961 in an area roughly 7 miles across and centering on Sable Pass. The family had not been seen in 1960 when the mother was followed by spring cubs. In 1962 the family was seen 10 times in the Sable Pass area from 19 May to 12 September, over an area 7 miles in diameter. Thus it was obvious that the year the mother was followed by her spring cub she occupied a range different from the one used the following 2 years.

Between 25 June and 26 August 1961, a blond mother and two 2-year-old cubs were seen 26 times on Sable Pass in an area about 4 miles across. The cubs had separated from the mother on 26 August. This family escaped my observation the 2 previous years, so apparently there had been a shift of range during the third summer that the cubs were with the mother.

A mother and her spring cub were seen 19 times during the summer and fall months of 1940—from 5 June to 9 October. When last seen, they were on Sable Pass wading in snow a foot deep. Their range centered on Sable Pass but they wandered west of East Fork River at least once and moved down Igloo Creek to Igloo Mountain. On 1 August I saw them feeding on berries along Igloo Creek for a distance of over 2 miles to the slopes of Igloo Mountain. The range over which they were seen was about 13 miles. On 25 May 1941 I saw the family on the bars of East Fork River. I remained in the park until August but did not see the family again so it seems the mother made a definite shift in her home range in 1941 (I was absent from the park in 1942).

Families Seen in Same Area Two Successive Years: A dark blond female with two yearlings were observed seven times from 17 June to 31 July 1966, on the east and west branches of the Toklat River, in an area extending 4 or 5 miles. This may have been the same female seen the previous year with two spring cubs at the eastern edge of the 1966 range, but positive identification was not made. During most of June, and again on 16 July, 1967 this family was seen eight times in the eastern half of their 1966 range. The range of this family probably was greater than that observed by about 5 miles, because many sightings were far south of the road, and on several occasions they moved out of sight farther south.

On 2 August 1969 a light-colored female with one spring cub were seen on Igloo Mountain. They remained in this area until 6 August. When next seen, on 25 August, they were at Tattler Creek and on 31 August they were near Igloo Creek at the east side of Sable Pass. This family moved nearly 5 miles during the month. Again in 1970, this family was first spotted on Igloo Mountain on 15 June and later on 19 June. Four days later it had moved to Sable Pass, but was not seen thereafter (I left the park early in 1970, my last summer observing bears.)

Another blond female with a 2-year-old cub spent most of summer 1969, from 2 June to 15 August, between Sable Pass and Igloo Mountain, an area 6 miles in length. They were seen eight times near the east end of Sable Pass during June and the first half of July. On 28 July and thereafter, they were seen eight times from Milepost 34 to Milepost 36, just east of Igloo Mountain. On 5 June 1970 the family was reported at Milepost 35 still together.

Some Additional Families Seen in an Area for Two or Three Months: The mother and one of three spring cubs (two of the cubs killed by another mother 10 July) were seen on Sable Pass 28 times at short intervals from 15 June to 21 September 1950. Their total wanderings during this period appeared to be confined to an area about 4 miles long by 2 miles wide. This family was not seen in 1951. Either it shifted its home range or the mother had lost her remaining cub.

A mother and two spring cubs were seen on Sable Pass 17 times between 26 June and 23 September 1950. Their observed range was about 7 miles by 2½ miles. This was the mother that killed two of the cubs belonging to the mother mentioned above. The family was seen in the same area nine times between 28 June and 15 September 1951. (I was absent from the park when the cubs were 2-year-olds.)

Between 11 July and 15 September 1960 a mother and a spring cub were seen six times in an area about 2 miles long at Highway Pass. Others also reported the family in the area during the summer. The family also was seen in the area on 6 June and 2 and 14 September 1961. The country was rough and broken and I seldom visited it, so that many sightings could not be expected. Apparently, the family was ranging in the same vicinity both summers. It was not seen in 1962.

A mother with two yearlings, seen in 1959, seemed more mobile than most families. I saw this group 14 times between 30 May and 31 August. On 30 May it was busily occupied digging roots on Polychrome Pass. The following day I saw it traveling on Sable Pass, 6 miles from where it had been feeding on Polychrome Pass. A few days later it had moved 2 miles to Tattler Creek to dig roots. In August it was seen near the East Fork River on two occasions. Thus it ranged over an area about 10 by

5 miles. I was not in the park when the cubs were spring cubs. The family was not seen in 1960 when the cubs were 2-year-olds.

During the summer of 1969, four families were observed repeatedly throughout the summer within a definite, fairly localized area. A female with two yearlings were seen 19 times from 28 May to 18 August on the west branch of East Fork River in an area about 3 miles long and 2 to 3 miles wide. They apparently did not stray much, if at all, from this vicinity throughout the period of observation.

A female with two spring cubs, one of which was lost in early June, spent all summer (26 May to 1 September) in an area centered on Sable Pass where 21 of the 44 sightings occurred. The family was spotted first on the slopes of Cathedral Mountain, and after moving to Sable Pass on 11 June, it returned to Cathedral on three occasions. In the middle of August the female and one remaining cub moved west to the bar of East Fork River and remained there for the rest of the summer. The extent of this summer range was at least 7 miles long.

Another family, female and one yearling, also spent the major part of the summer on Sable Pass in 1969. Between 29 May and 29 August the family was seen 32 times, 25 of these observations on Sable Pass. The family spent a few days on the East Fork River bar in early June and again in August, and twice in July moved down Igloo Creek about a mile. The length of the area known to be used during the summer was about 5 miles, but most of June and July were spent within an area of 1 or 2 square miles on Sable Pass.

A fourth female, followed by two spring cubs, was seen 22 times, from 9 June to 1 September, in an area about 9 miles long between Highway and Thorofare passes. The family did not remain long at any spot, but ranged widely back and forth over its summer range throughout the period it was observed.

None of these families was identified with certainty in 1970 when I was in the park for only the month of June, but single sightings of families that I and others made probably included at least two of them.

Home Range of Males

Adult males wander widely, especially during the breeding season when they are seeking mates. One day a male was seen in the morning traveling steadily up Igloo Creek. When I saw him in the late evening, he was traveling steadily down East Fork River, going out of sight around a bend. Where he had gone in the Sable Pass area I do not know but, in a direct line over the pass, he had covered 7 or 8 miles. This male was not seen again.

On 13 June 1959 a crippled male appeared on Sable Pass. Between 20 June and 10 July he kept company with two females, mostly in an area a mile or two across. After 10 July, he was alone and fed in the general vicinity until 26 July, the last day I saw him in 1959. Reliable observers stated that they had seen this male in the area in 1958. During 1960, 1961, and 1962 this male was not seen. But from 11 to 17 June 1963 he kept company with a female on East Fork River and part way up Sable Pass. After this date, he disappeared. On 7 June 1965 he was observed on East Fork River, where he had been seen in 1963. He was seen in 4 different years over a span of eight summers. His range was wide enough so that he escaped being seen during 4 of the years he was known to be active.

In his notes Olaus J. Murie tells about the wanderings of a male grizzly whose tracks he followed in the snow on 26 September 1921. Olaus was near Circle, Alaska, observing caribou. He came upon a grizzly track and followed it over the hill to a caribou that had been shot some time before: "He [the grizzly] had carried the meat down the hill into the woods, returned up the hill, then wandered westward and turned down in the woods again."

The following day Olaus again followed the track which was partly snowed over.

> It led me a long chase. First it went down the hill into Smith Creek where he had walked in the water for about 100 yards, then up the opposite hill through thick timber and willows, over the bare ridge, then down into the next creek bottom. Here he had walked in the water about 300 yards upstream. The tracks led up to the head of the stream, through a mass of willows and small spruce, over a low saddle and down the other slope. He had angled off on to a ridge and followed the bare ridge toward a mountain, then down into another wooded creek slope. Here he finally entered a thick growth of spruce and I noticed he wandered back and forth and circled around and I found two places where he had dug in the moss a little. I went very carefully then, but presently I came to his bed, from which he had just fled! He had scraped out a hollow in the ground over a foot deep, and there he had been lying. He had become very frightened, for the tracks showed that he ran in long leaps, knocking over rotten stumps and small spruces as he went. He apparently injured a toe of the right hind foot, as indicated by an occasional blood spot. He fled in the direction of my camp, which was fortunate, as it was late and I had all I could do to reach it before dark. As he ran, the bear kept to the timber and I did not get a glimpse of him. I finally left the trail and went to the camp. The hind track measured 10½ inches, including claws, and the front track was 7 inches wide. Altogether the bear had travelled 6 or 7 miles in his wanderings (Murie's notes 1921).

Home Range of Lone Bears

Various lone bears were observed and identified over periods of 1 or 2 months, but for longer periods identification usually was uncertain. Bears, if they take the notion, may wander far in a short time. A bear that visited various camps was known to move over 20 miles in one day. A young bear, crippled on a front foot, was seen on 2 successive days and had traveled 6 or 7 miles when seen the second time.

On 30 August 1959 at Milepost 67 I saw a small bear, perhaps a 3-year-old, in a new cream-colored coat. Its face was brown, its legs blackish. This is by far the lightest colored grizzly I ever saw. It had the appearance of a female. On 4 June 1960 I saw this same bear feeding on a carcass of calf caribou on Highway Pass (about Milepost 57). It was seen on Highway Pass again on 19 June, and at Milepost 56 on 5 and 22 August. In 1961 it was seen on Highway Pass on 6 and 10 July, and at Milepost 66 on 17 August. This well-marked bear was seen over an area about 11 miles across between 30 August 1959 and 17 August 1961.

Home Range of Twin Cubs

When cubs leave their mothers and wander about on their own, I expect there could be either considerable dispersal or the cubs might continue their wanderings in the pattern of their mother with whom they had traveled for two full summers and part of a third. Data were collected on one pair of cubs over a period of three summers after they had separated from their mother. Their movements, so far as is known, followed the pattern of their mother.

In 1959 the two cubs were yearlings and followed the mother closely. The family was seen 31 times in the Sable Pass area between 17 June and 2 August, in an area about 6 miles across. In 1960 mother and cubs made their first appearance on 18 May. Two days later the cubs were alone. They spent the summer of 1960 in the Sable Pass area, covering about the same stretch of country as they had the previous summer. One or both bears were seen on 53 days in 1960, between 20 May, when they were first seen alone, and 26 September.

I first saw the two cubs, now 3-year-olds, on 9 May 1961 on Igloo Creek, a mile from where they first were seen in 1960. They were observed in the Sable Pass area from 9 May to 18 September; one or both bears were seen on 52 days.

In 1962 one of the cubs was seen first on 17 May and the other on 22 May, again on Igloo Creek near where they were first seen in 1961. One or both cubs were seen in the Sable Pass area on 23 days between 17 May and 23 August. They moved away from their usual summer haunts earlier in 1962 than they had the 2 previous years. Possibly they followed somewhat their mother's pattern and moved to lower country in search of berries. These cubs were not recognized in 1963 when they were 5 years old. A more detailed discussion of the movements of these cubs is given in the section dealing with cub companionship.

The home range of the two cubs during the periods in which they were observed covered an area 9 or 10 miles in diameter. The cubs, when on their own, did not follow their mother's usual home-range pattern of leaving Sable Pass near the start of the berry season, but stayed on into autumn the first 2 years and quite late the third season. They also appeared much earlier in the spring in the Sable Pass area than did their mother.

In 1963 two other cubs, apparently 2-year-olds on their own, were seen on five occasions between 9 and 27 September. During this period they ranged from Savage River to Teklanika River, a distance of about 9 or 10 miles. They were feeding chiefly on berries, but finding them scarce probably wandered in a wide search for more.

In 1963 a pair of cubs that appeared to be 2-year-olds was seen near the top of Sable Pass on 10 occasions between 23 July and 10 August.

Many other young bears were seen frequently over a period of a month or two, but their identity was not closely maintained for longer periods so data for them will not be tabulated. One year, for example, seven or eight young bears, from 2 to perhaps 4 years old, roamed over the Sable Pass area for much of the summer.

Summary—Home Range

Three families were recognized for 4 consecutive years, two with a single cub each that was still with the female as a 3-year-old, and one seen for each of 2 years with yearling, then with a 2-year-old, and subsequently with another litter of cubs. Data were secured over a 3-year period for eight females while they were followed by cubs. Twenty-seven other families were seen in 2 consecutive years; my absence from the park in the year the cubs were born or when they were 2-year-olds prevented additional records for 15 of these females. Of the other 12 females, 7 were not seen with 2-year-old cubs and 5 were not seen with spring cubs. The usual separation of these cubs from their mothers, in one case as early as 20 May, makes the opportunity to see a 2-year-old with its mother unlikely. Home-range information on all these families seen in more than 1 year is summarized in Table 5. Home-range data for a number of the 69 families seen more than once during a single year only are presented.

The data on home range show that grizzlies have a strong tendency to use definite ranges of limited extent year after year. Each bear tends to follow its own pattern of movement. For instance, some families spend most of their time in one general area throughout the root-, grass-, herb-, and berry-eating periods, whereas other bears may remain in this same area for only the grass- and herb-eating periods, and feed on roots and berries in adjoining areas. Home ranges of different bears in an area thus overlap in various ways.

The home-range pattern for some families was similar from year to year, whereas others varied their movements.

The observed home ranges, over periods up to 3 or 4 months, generally varied from 5 or 6 miles to 12 or 13 miles. However, more extended movements are known to occur; over a 4-year span one family ranged at least 22 miles. For periods of a few weeks, bears may confine their movements within an area a mile or two in diameter.

For none of the families or single bears was the *total* home range known. Within the ranges of the bears observed, dens were scattered widely, which suggests that bears often did not travel great distances from the den.

General observations indicate that adult males wander more widely than do females with cubs, at least during the period when the males are searching for mates. Because recognition of males and other lone bears is more difficult, documentation of their home ranges was rarely possible.

The varied habitat over most of the park makes spring, summer, and autumn foods available over limited areas. Extended movements, greater than seem necessary, may be made nevertheless. In years when berry crops fail (which are rare), bears may wander more extensively although such movements usually are similar to the home ranges in other years.

Joint Occupation of Range

Grizzlies assume no private ownership of territory. Joint occupation of ranges prevails and bears wander freely over the countryside. Smaller bears keep out of the way of the bigger bears as much as possible. Each unit, such as the lone bear, breeding pair, mother and cubs, sets of older cubs on their own, is independent and does not fraternize ordinarily with other units. When bears do feed within 2 or 3 hundred yards of each other, where they have been attracted by good rooting or grazing, a certain amount of uneasiness and watchfulness prevails, the degree of anxiety depending upon the types of bear units that are present and perhaps the extent of previous acquaintance.

To some extent a peck order exists—each bear knows fairly well where it belongs in the hierarchy. Its status may not be determined by conflict but by recognition of the class to which it belongs. Each bear knows which classes to fear, to defy, and to dominate or tolerate. For example, a female bear will maintain distance from a large male, perhaps defy another female, and, to a degree, dominate or tolerate young bears not yet full grown. Bears are long-lived which gives them time to become familiar with one another, and as acquaintance and experience increases, the peck order probably becomes more individualized. In a wilderness such as McKinley National Park, however, many of the associations are too distant to become very personal. But when two males seek the same female, for instance, dominance between them is settled by bluff or perhaps by conflict. A bear in possession of a carcass must decide his status in relation to an intruder; if both bears feel equal to the other there is a showdown and status is determined by a scuffle.

When choice food is available in a restricted area, a special tolerance may develop. Near the Alaskan coast a number of bears may be attracted to a limited stretch of water where salmon congregate during the spawning season. The lure of delicious food decreases their timidity, and as the stronger bears become accustomed to the proximity of others, their intolerance decreases, and a sort of truce develops although some degree of intolerance usually remains. The most extreme example of this is at a garbage dump, such as at Yellowstone National Park, where I have seen grizzlies, side by side, wallow degradingly in the garbage.

In McKinley National Park, bears ordinarily do not congregate in limited areas, but in some years a fairly high density of bears occurs in an area about 5 or 6 miles in diameter on Sable Pass. In 1961, 5 families, 2 sets of twin cubs 3 or 4 years of age, and at least 6 lone bears (a total of 23 bears) were seen throughout most of the summer on Sable Pass. Fifteen of these animals were seen on 1 September along a 7-mile stretch of road. A total of 28 bears, including five families, occupied the Sable Pass area for much of the summer of 1962. Yet no notable conflicts among bears were seen in 1961 or 1962 despite this concentration. In another area, between Mileposts 24 and 36, five families were seen along a 2-mile stretch of road during 1 week in August 1969. This was a favored blueberry spot and the beginning of the berry season coincided with this brief concentration of bears.

When bears become aware of each other in their travels or feeding activities, there is a mutual appraisal which at times seems to be quite rapid. Status depends chiefly on size. The first reaction of both parties is to move apart. The bear most startled may make the first move away and thus obviate a similar reaction on the part of the other, who may stand and watch or give a perfunctory, brief chase. Young bears are always ready to retreat and so are mothers with cubs, except when they recognize the other party as a young bear. If a big male encounters a smaller bear, he recognizes his own superiority and may either disregard the other bear or, if quite close, may make a token run in its direction. The smaller bear makes his own evaluation and if the other is too near for comfort, he hurries away. There are, of course, endless variations, behavior depending much on past as well as immediate circumstances, including the degree of familiarity.

In general, one discovers a bear or a family off by itself, comfortably apart, feeding at ease. But in some favorite feeding areas, such as on Sable Pass, one often finds that two or more bears are, by chance, feeding quite close to each other. They seem oblivious sometimes, but at other times are gauging the safety of the situation, keeping aware of the other bear's position and moving accordingly, making a fine adjustment or departing from the neighborhood.

The following incidents illustrate various kinds of behavior when bears get involved with one another.

A Mother Beyond Her Usual Range

Individuals, man or beast, are more confident when on familiar territory. A man in his own home tends to speak with more composure and confidence than he does out in company. At our winter bird-feeding board, the red squirrel living in a nearby cabin acts with authority, chasing with vigor magpies, jays, or visiting squirrels. On the other hand, this squirrel, when visiting elsewhere, behaves like an intruder, is meek, tolerates all the birds, and usually hurries homeward lest he chance to meet the squirrel in charge.

Most grizzlies I have observed in McKinley are, so far as I know, on familiar ground and are not beset by the added worry of being in a strange area. On a few occasions I saw mother bears that had ventured beyond their familiar ranges. These mothers were excessively alert and apprehensive. On 22 August 1962 I watched a dark female with her two spring cubs for several hours on the flats of Polychrome Pass where they had come to feed on buffaloberry. This was, so far as I know, her first appearance in the area during summer. Apparently, she had wandered over from the broken country to the north. She behaved as one would expect a bear to behave in strange country. She seemed to be on edge, was ever watchful, and received several false scares. When, later in the day, she spied a young bear about 400 yards away, she hurried away immediately without trying to get a better look.

Three Families and Three Lone Bears on Sable Pass

Early in the morning of 10 July 1959, two females, each with twin yearlings, were feeding on the west side of Sable Pass, some two-thirds of a mile apart. Although Sable Pass is above timberline, bears, even in close proximity, may not always see each other because of the draws, depressions, hummocks, and patches and strips of tall willow brush. A third mother, followed by two yearlings, passed southward on top of the pass a short distance to one side of the first two families, and continued to move toward a pair of mated bears a mile away. A lone bear was present on the west side of the pass, hidden most of the time from all the others. The three families, the pair, and the lone bear were all within an area about 1½ miles in diameter. These bears, spending much of their time in an area 5 or 6 miles in diameter, were not usually so concentrated.

The first two families fed on green vegetation in the moist hollows and swales, inadvertently hidden from each other. A brown female to leeward, however, knew that the other family was somewhere upwind. Both she and the cubs raised their muzzles occasionally to better test the breeze carrying the scent. I expect that this brown female knew who was feeding in the hummocks to the south and knew that it was the golden female with the two golden cubs, for this was not their first meeting. She was alert, but being accustomed to scenting other bears and especially the golden bears, while feeding in this area, did not hurry away.

At noon, the golden family fed downward among the hummocks to about one-quarter mile of the brown female family. After a time the brown family moved a little nearer and the golden mother became aware suddenly of the brown one as the latter came out of a small depression. The golden female and her cubs were surprised, having been unaware of the presence of another bear, and galloped back up the slope, the cubs in the lead. After retreating 300 yards, they regained their composure and fed slowly upward. The brown female, not so startled, watched the family flee and resumed feeding—the situation thus resolved by the flight of the golden bears.

In the afternoon, the two families and three lone bears (the mated pair had separated) were all feeding on the west side of Sable Pass. They were evenly dispersed and behaved as though unaware of one another. At 5:00 p.m. the golden bears, again following choice grazing down the slope, were feeding 200 yards from the brown family which had fed in a small area all day. When the brown female moved to a little rise, she was discovered and once more, led by the two yearlings, the golden family galloped up the slope. At intervals, the cubs stood on hind legs for a better look, and after each look, galloped on. The golden bears hurried away in this manner for 600 yards, then stood watching for some time, made another gallop, and settled down to a steady walk, now led by the mother, until a half-mile away when they disappeared over the horizon. The mother, one would guess, felt the area was too congested and preferred to move away.

The brown female moved a short way toward the hummocks where the golden family had been, sniffing the air as she advanced. She then resumed feeding but soon moved off toward the top of the pass as though she too thought the place was getting crowded. At 8:00 p.m. I discovered the third family (last seen at noon) on the west side of Sable Pass. When the mother saw one of the three lone bears 300 yards away, she and her cubs galloped up the slope a short distance, then settled down to a steady walk that took them out of sight a half-mile away. These families had some acquaintance and tolerance but preferred to retreat from the annoyance and worry of nearby company. This behavior suggests that a certain amount of tension can be endured for a time, but then the bear seeks relief by moving away.

On a few other occasions the three families all appeared at one time, but they were spaced widely and usually only one or two families were in sight at any one time. Without special effort, they seemed to keep apart even though their wanderings, as noted above, were mostly confined to a jointly occupied area 5 or 6 miles in diameter.

Brown Mother Retreats from Blond Family

On 24 June 1962, 3 years after the above incident, the same brown female and her two yearlings (another set of cubs) made their first appearance of the year on Sable Pass. She had spent at least the previous three summers in the area and was on home ground. After feeding for a time on crowberries, the bears crossed over into a hummocky area to feed on the new green vegetation. The depressions between large hummocks and the scattered patches of willow brush were such that a bear was fairly well hidden from any other bears in the vicinity.

Some distance up the gentle, uneven slope, about one-quarter mile from the brown female, a large blond female grazed with her two small yearlings. She lay down and nursed her cubs, then lay resting for a half-hour before feeding slowly toward the other family. Neither family was aware of the presence of the other until the blond mother approached to within 150 yards of the brown female. At this point the latter, who was downwind, scented the blond family, raised her nose a few times to test the air, then stood on hind legs to look. When she saw the blond mother, she dropped on all fours and galloped away for 150 yards. Here she and the cubs were apparently at ease, for they began to graze. The blond mother and two cubs stood erect on hind legs to watch the other family gallop away, then resumed feeding. For the next 2 hours the two families fed about 300 yards apart.

The adults and cubs in each family were always conscious of the proximity of the other family. Occasionally, they watched each other. The brown mother, who had been the most startled and made the initial retreat, seemed to be the more nervous, and she finally moved away. The blond mother, because of the initial retreat of the other family or because of the outcome of earlier, unobserved encounters, probably had a psychological advantage and tended to be more composed. The brown mother's behavior was the reverse of what it had been in a similar incident 3 years before.

Two Mothers of Spring Cubs Avoid Proximity

On 29 June 1964, a blond mother with a spring cub crossed a snowfield on Sable Pass and grazed upward on the opposite slope toward a mother grazing with two spring cubs. When the second female sighted the family down the slope, she galloped 300 yards away and at once nursed her two cubs, possibly displacement activity brought on by her nervousness. After the brief nursing, the mother started grazing again. Soon, two trotting caribou startled her and a little later she was in the line of travel of 60 caribou. Other caribou in migration were passing by, singly or in small groups, and the bear was kept on edge because she had to identify each moving object to make sure it was not another bear. In time she became accustomed to the activity of the caribou, regained some composure, and fed rather steadily among the hummocks.

The mother with the lone cub was not aware of the retreat or presence of the other family. She continued to feed among the hummocks, gradually approaching the position of the family above her. When she had the family in sight, but still quite far off, she retreated down the slope back the way she had come, and started up the snowfield she had crossed earlier. A lone bear over to one side caused her to change her course slightly. The cub, apparently oblivious of the reason for the retreat, amused himself by following the old trail across the snow, sniffing at each track. This caused delay and added to the mother's anxiety. She had to stop and wait, and for a time sat up and surveyed the country. When the cub finally left the old trail and veered toward the mother, she continued steadily on her way to the west. Later in the summer this family was seen, but not on Sable Pass. Their summer range the following year was still to the west. Apparently, Sable Pass was out of their usual range and may have been one reason for the long retreat. The mother with the two cubs remained feeding on the slope.

The same day two lone, young bears on Sable Pass came near each other in the course of their grazing. When the smaller one discovered the other, he galloped for a half-mile and continued walking away. The larger one returned to his grazing. The small bear was especially cautious.

Apprehension in Young Bears

The sudden appearance of another bear nearby is cause for considerable alarm for any bear. One day as I watched a pair of twins about 3 or 4 years old, one of them walked over the crest of a ridge in the course of feeding. A few minutes later, when he returned to view, his companion was so frightened that he galloped away. The mistake was soon recognized and the startled bear joined its companion for some reassuring play.

On 2 June 1960 I saw a small, blond bear, that I judged to be about 3 years old, digging roots near the base of a high bank along the Teklanika River. In the woods above the bear a moose came into my sight, feeding on willow. The bear heard the moose and stood up to look toward the sound, then dashed out on the bar about 50 yards and looked again. He could now see the moose and, after watching it briefly, returned to his digging under the high bank. The previous day I had seen another larger bear chase this one. When he heard the moose above him, he apparently was concerned to learn if another bear was approaching.

Young Bears Chasing Young Bears

Young bears, usually tolerant of each other, occasionally chase one another, sometimes perhaps in the spirit of play, sometimes apparently in an antagonistic mood. Once when a young bear wandered within 400 yards of another, he was chased about a half-mile. The one escaping was the smaller of the two. The large one returned part way to where he had been and began to feed; the other continued moving away from the area.

On 29 August 1964, I saw a young bear chase another up the East Fork River for about a mile. The bear being chased stopped and the other passed it, 100 or 200 yards to one side, and moved into the willow brush. The bear left on the river bar started to dig roots. This chase appeared to be a casual affair—a half-hearted, get-out-of-my-way action and possibly the spirit of play was somewhat involved.

On 2 July 1964, Zack Price saw a young bear chase a smaller one on Sable Pass for 12 minutes over an area about 1½ miles in diameter. He said that the bear in the lead seemed to gallop just fast enough not to be overtaken, and the one behind just fast enough not to overtake. This may have been play activity.

Injured Young Bear

On 18 September 1964, I watched a small dark bear on Sable Pass feeding rapidly and nervously on crowberry. Up high on his rear right hindleg was a fresh wound, and a piece of hide about 2½ inches wide and 7 inches long was hanging loose. Earlier in the day, about a quarter-mile from this bear, I had seen a blond bear that seemed to be somewhat larger. Both bears were 3- or 4-year-olds. Possibly this blond bear had inflicted the wound. The behavior of thc wounded bear indicated that the altercation was recent, because at short intervals it would stop feeding for a quick look around. The joint occupation of range apparently had caused an altercation. The bear was oblivious to my proximity.

Two 2-Year-Olds Chased by Male

Small bears, those from 2 to 4 or 5 years old, apparently can, and know they can, outrun a large bear, especially in steep terrain. Once I watched two 2-year-old cubs (recently separated from their mother) discover a large male and gallop across a wide river bar to a steep slope. The male loped along far behind, moving more slowly. On the mountain slope the two cubs stopped at intervals to watch the labored progress of the puffing male who had to stop frequently to rest and catch his breath. When the young bears reached the ridge top, they moved out of sight; when the male reached the top, he gave up the chase. Other species also seem to gauge their margin of safety or their degree of vulnerability. One fall I watched a silver fox behave much like the two cub bears when it was chased up a mountainside by a coyote. The fox stopped on prominent rock ledges to look down and bark at the coyote whose climbing was much more labored. Dall sheep seem especially aware of their advantage when they are in rocks above a bear or a wolf.

Two Young Bears Chase Family in Play

On 25 August 1962, on the east slope of Igloo Mountain, a mother and two yearlings moved down a low ridge into a brushy swale as they foraged. Behind them on the crest of the ridge, twin bears, 3 or 4 years old, appeared. The two young bears stood erect on their hind legs for a better view of the family below them. The two yearling cubs, upon seeing the two young bears on the skyline, took fright and dashed downward, the mother following. When the two young bears saw the family fleeing, they caught the spirit of the chase and loped down the slope after them, but when the mother stopped, they put on the brakes, and as she started a deliberate walk toward them, they retreated at a lope. When the mother turned and walked toward her yearlings, the two young bears again pursued at an easy lope. Seeing the mother stop, the young bears also stopped, but when she again started for her yearlings, they followed. The female's patience had come to an end; she turned and charged as though she meant to follow through. She chased the two youngsters to the top of the ridge where they stopped when she turned to hurry back to her yearlings. The two stood briefly, watching the family as it moved down the slope, then gave up the game and retreated down the opposite side of the ridge. They were aware of their inferior status, but apparently had confidence in their ability to escape and made a game of it. Perhaps they also knew from previous experience that a mother usually will not carry on a chase very far from her offspring.

On several occasions young bears were seen near families without any member showing much concern. If the youngsters kept to a respectable distance, they were tolerated.

Two Self-Assured Young Bears

I once saw two 3-year-old bears cross the line of travel of a mother and her two spring cubs. As the young bears moved along in a dignified, grownup manner, they obviously were watching their distance relationship to the family. They crossed about 150 yards in front of the family that was coming up the slope. About 100 yards off to one side, they stopped and sat down to watch. The mother with her spring cubs veered slightly toward the two young bears who took the hint and started walking away. The mother resumed her original course and seemed unperturbed. She had threatened only sufficiently to let it be known that she would not tolerate bothersome familiarity, and the two youngsters, rambling about on their own, seemed to understand.

Bluff by a Mother Bear Fails

There are occasions when a mother bear, feeling superior to a large 3- or 4-year-old animal, resents his presence and walks toward him to chase him away. On two occasions I have seen a large, young bear that appeared to be a male stand his ground and refuse to be bluffed. On both occasions the female discreetly returned to her cubs rather than risk an encounter with a bear her own size.

Families Unperturbed Near Young Bears

On 26 July 1963, I saw a female with her two 2-year-old cubs feeding on Sable Pass about 200 yards from one lone bear and 300 yards from a second lone bear. These lone bears were 3 or 4 years old. Later, the family moved within 75 yards of one bear that was feeding in a hollow directly below. The mother and two cubs stood watching the young bear who was too busy feeding to notice. But the family, unperturbed, moved away, feeding. Later, the second young bear moved within 75 yards of the family, apparently neither unit knowing the proximity of the other. When the mother discovered the young bear, a very shaggy animal, she lay on her stomach for 5 or 6 minutes with head resting on paws, watching it. Her two cubs stopped feeding, walked to her, and one pushed its muzzle under her chest. She complied and rolled over on her back, out of sight of the unaware, shaggy bear below them, and the cubs nursed. In watching the young bear below her, she apparently did not like to have it feeding so close, yet was not concerned or resentful enough to do anything about it; that is, either to chase it or to retreat. Nursing finished, she sat up for a look; then she and her cubs lay in a heap. The shaggy bear gave no indication that he was aware of this family. He did not catch their scent and fed too steadily to see them when they were in view. The two lone bears, now both on the slope below the family, rested while the family moved on without disturbing them.

On this day I saw, in an area about 4 miles in diameter, the above mentioned family and two lone bears and in addition seven other young bears. The latter group consisted of twins that were 3 or 4 years old, three lone young ones, and twin 2-year-olds on their own. These five groups of bears were, for the most part, well spaced during the day. The 2-year-old twins, which were quite timid, at one point galloped away from one of the lone bears that had chased them a short distance. The area was much more congested than usual. The above situation is described briefly to show again that when a mother with cubs recognizes young bears nearby, she may ignore them to some extent.

Young Bear Shows No Fear

On 23 July 1965, on the low pass south of Cathedral Mountain, I saw a mother with two yearlings on the far slope of a hollow. After a time, a blond bear, definitely smaller than the mother bear, appeared out of a draw and came in sight of the family on the opposite side of the hollow. The lone bear, on seeing the family, seemed to take on a stiff, guarded gait, and the family, startled, moved back 15 or 20 yards. Then the mother started walking slowly toward the lone one which was now 75 yards or less from her. The small bear stopped, walked forward a few steps, and stopped again. The female, followed closely by her cubs, started galloping toward the small bear and disappeared in the hollow out of my view. I expected her to appear momentarily on the rise where the small bear was standing, but instead the two cubs reappeared and galloped in retreat back toward where they had started, followed a moment later by their mother. The cubs probably caused the retreat. They stopped soon after coming into view; the small bear started to walk away, and the family resumed feeding.

A Mother With Spring Cubs Tolerant of Young Bears

On 3 July 1962 a mother and two spring cubs on the middle of an old river bar near the head of the East Fork River were grazing on peavine which grew there luxuriantly. Some young bears had also been attracted to this peavine. About 250 yards to one side of her, at the edge of the bar, were twins about 3 years old. On the other side at a similar distance from her was another set of twins about the same age. These three sets of bears, brought together by choice grazing, were adjusted to each other's presence and fed or rested in relatively close proximity, with toleration and composure. The female had no fear of the young bears, and they in turn apparently had confidence that the female only wished to be left alone and that they could escape if she should give chase. They adjusted distances accordingly, distances which no doubt shortened somewhat with familiarity.

A Mother Chases Young Bear for Long Distance

On 12 August 1959, a photographer saw a mother bear chase a small, lone bear (about a 3-year-old) across an extensive talus slope on Igloo Mountain. The distance traveled was a little over one-half mile. At one point the mother almost overtook the small bear. When the chase started, the mother's two yearlings climbed far up among cliffs, away from the action. As I arrived on the scene the chase was over, the 3-year-old was in the distance, moving away and the mother had moved up in the cliffs to retrieve her cubs.

A Mother Bear Causes a Larger One to Leave

On 3 June 1964, on Cathedral Mountain I saw a mother and 2-year-old cub moving slowly up a slope as they dug roots. A few hundred yards higher on the mountain and on an adjoining ridge was a rather large bear. About 2½ hours after I first saw these bears, the mother and cub, which had been out of view for a time, reappeared not far below the large, lone bear. The mother started to walk toward the lone one which was obviously larger then she. The big bear moved away but stopped two or three times only 25 or 30 yards ahead of the mother. The cub remained far behind. The mother stopped and the larger bear moved 200 or 300 yards across a ridge. The mother's behavior was different from anything I had observed previously. Ordinarily, instead of advancing toward the lone bear, a mother would have retreated.

Large Male Frightens Mother With Spring Cubs

On 13 June 1959, an old, crippled male, traveling the country in search of a mate, came down a slope and surprised a mother with spring cubs feeding along a streamlet. She was not aware of the male until he was 50 yards from her. She and her cubs galloped away and the startled male, after a brief look, made a short, token chase of a few yards before continuing on his way down the slope. The family hurried on until it reached the top of a ridge a half mile away and, without stopping to look back, went over to the other side. The mother must have recognized the male as a big one, and lost no time moving away. It is probably to avoid big males that females with cubs seek cliffs for a night bed in spring and early summer.

Large Male Unsettles Mother Bear

Early in the morning of 8 July 1965, a mother and her 2-year-old cub were seen feeding on an extensive old river bar covered with sod and scattered patches of willow brush. They were still there at 6:15 p.m. as was a large male on the bar 300 yards away, all grazing steadily. At 7 p.m. the cub stopped feeding, rubbed his back against willow brush, sat beside his mother, and nursed. A half-hour later the mother discovered the male although he did not see her. She and the cub galloped out on the gravel bar, walked away in a large arc, and returned to the green bar farther to the north, a quarter-mile from the big male who was still grazing steadily. The mother lay down but stood again when the cub crowded her for a nursing. I then noticed two black wolves trotting briskly across the gravel bar toward the two bears. When one passed a few yards to one side, the mother made a few jumps toward it. The cub made a short gallop toward the wolf passing on the opposite side. The wolves continued northward. About 250 yards down the river, one of the wolves turned and trotted back toward the family, into the wind. When the bears discovered the wolf some 50 yards away, they were startled, and without taking a good look, galloped away. It is probable that the mother, aware of the big male grizzly in the area, was especially wary, and on seeing a black animal approaching, partially hidden by scattered willow brush, assumed it to be the male, and departed. The family settled down to a steady walk and continued for over a mile before

Fig. 19. A female and her spring cub watch intently as a lone bear passes nearby.

feeding. This seems a good example of the behavior of a bear being influenced by a recent experience (Fig. 19).

A Male Chases a Cub

Large male bears may attack smaller bears, including cubs. Apparently, it is the big male that is especially dangerous in this regard. On 10 July 1961, a group of hikers far up the East Fork River on a bank about 15 feet above the river bar, saw what appeared to be a large, male grizzly. As he moved along, he stopped occasionally to listen and test the air. Then they saw him begin to gallop, and a mother and yearling appeared in front of him, running away. The female turned to ward off the male who continued in pursuit of the cub. The mother nipped at the male from behind, causing him to turn on her twice. The chase led directly under the bank where the hikers stood. As the cub turned up a draw, the male continued forward along the bar and was last seen a mile away. The presence of the hikers may have disrupted the chase or perhaps the female's attacks on the rear of the male were effective. The attacking bear was considered a male because it was much bigger than the female bear.

Deaths at Garbage Dumps

On 28 August 1963, the mother of three spring cubs was found dead near the park garbage dump. She had frequented the dump for a few weeks, but I did not learn about the incident until the evening of 29 August. When I arrived at the dump that evening, I examined the carcass which had been dragged a short distance to the edge of the dump pit. There were deep tooth wounds on the head and neck; teeth had grazed and penetrated the skull and the wounds were bloodshot, suggesting that they had been made while the female was still alive. It is possible that she had been attacked by a large male while she was trying to protect her cubs. This is suggested by the behavior of the mother described in the preceding episode. The teeth were quite worn, the molars, down to the gums, indicating that the female was very old. If she did have an encounter with a large bear, it is likely that she lost some of her maneuverability in fighting, and this prevented her from escaping the male.

At 1:30 a.m. on 23 July 1961, an archeologist came upon a large bear straddling an inert bear on the highway at Milepost 6, near the park garbage dump. The big bear's jaws were clamped on the neck of the victim as he dragged it across the road. At intervals, he would lift the inert bear and give it a shake, as though he had just attacked and killed it. A hotel employee reported that earlier he had seen two bears in the area playing. This "playing" probably was a serious altercation resulting in the death of one animal, the one seen by the archeologist. When I examined the carcass of the bear later that day, the inguinal region and the flesh on the lower ribs were eaten. On the ground under the carcass there was much blood from neck wounds. The sex of the bear was not determined. What we saw indicated that a large bear had killed another whose age we estimated to be about 4 years. I had often seen a light-colored bear, resembling this one, at the garbage dump, and it was not seen again after this incident.

Both these incidents appeared to result from congestion at a garbage dump, indicating that the tolerance among bears usually observed there can break down. Large grizzlies are known to have killed cubs at garbage dumps in Yellowstone National Park also (Craighead and Craighead 1967).

Tragedy as Result of Joint Occupation

Elsewhere (Murie 1961) I have written about two mother grizzlies who spent the summer of 1950 on Sable Pass. One, whom I called Nokomis, had three spring cubs; the other, called Old Rosy, had two spring cubs. Whenever they found themselves uncomfortably near each other, they moved apart. I saw Nokomis 28 times during the summer and Old Rosy 16 times. They confined much of their wandering to a small area a couple of miles in diameter. On 10 July a companion and I started to watch the families early in the morning; as we watched them moving about in their grazing, generally 300 to 400 yards apart, we wondered what would happen if the two families should meet accidentally at close quarters.

Late in the afternoon, about 5 p.m., Nokomis with her three cubs moved westward, passing below Old Rosy and her two cubs about 150 yards above. A short time after the mother and three cubs had disappeared in a hollow, one of the cubs came back over the trail and behaved as though he was lost. Old Rosy watched the cub from up the slope and then galloped toward it. The cub retraced its steps at a gallop, and Old Rosy followed, both disappearing in the hollow. Later, when we saw the bears, the three cubs were far ahead, climbing Sable Mountain and stopping at intervals to watch the two mothers. Their mother, Nokomis, kept intercepting Old Rosy who tried to get past her. On a steep slope they fought, with Old Rosy having the advantage of being above Nokomis when they clashed. They then continued up the slope, walking side by side for a time, and later Old Rosy galloped up toward the fleeing cubs who were nearing the top of Sable Mountain. Nokomis remained below as though she was not aware of the location or plight of her cubs. When Old Rosy overtook the cubs, one escaped downward, one upward, and she killed the third. She then went after the cub that had gone upward and apparently killed it because the next morning eagles were feeding on its carcass. In this instance, occupation of the same range by two families resulted in tragedy—a natural curb of the bear population.

A Potential Conflict

On Sable Pass on 10 July 1955, I witnessed what might have developed into another tragedy. Two families, a mother with one yearling and a mother with two spring cubs, fed toward each other near Igloo Creek. The spring-cub family was feeding in the open at the edge of a patch of tall willow brush. The other family was moving through the brush toward them, the yearling some distance in advance of its mother. When it was about 25 yards from the spring cubs, the yearling in the brush stood on its hind legs, apparently having scented the other family, and bawled three or four times. This caused its mother to hurry forward anxiously. Just before breaking into the open, the mother saw her cub over to one side and turned toward him. Otherwise, she would have burst into the open only a few yards from the spring cubs and a tragedy might have resulted. The spring-cub family, upwind, then became aware of the other two bears and galloped fast and far. The incident illustrates the manner in which families can meet accidentally at dangerously close quarters.

Grizzly Kills Two Black Bear Cubs

Isabelle and Sam Woolcock, who have spent several summers observing and photographing bears in McKinley National Park and are reliable observers, witnessed a tragic incident near Anchorage. Isabelle wrote me about it as follows:

> We were sitting on the river bank glassing the low mountain slopes when we saw a black bear with two cubs. As they came out on the grassy slopes they were plainly visible with the naked eye. Then we saw a lone grizzly coming over a rise, and as it travelled forward it was at intervals hidden by the brush. When it came into the opening where the family was feeding it made a short dash toward them, swatted one cub, and then the other as it attempted to run away. The mother seemed terrified as she dashed away. The grizzly then grazed slowly until out of sight.

I expect a big male may treat the cubs of a female grizzly similarly.

Grizzly Reported Killing a Black Bear

In O. J. Murie's field notes written at Ophir, Alaska, on 2 March 1922, I find the following item pertaining to bear conflict on a common range: "A trapper tells me that one fall before the freeze-up he found a black bear on Big River which had been killed by a brown bear. The black one had been digging a new den, near an old bear den, and had a bed at the base of a tree. There were marks of a desperate struggle. The body of the black bear had been deeply scored, apparently by the claws. A slight amount had been eaten. The black bear had been about half-covered by moss, etc. when the trapper appeared on the scene."

Mingling of Bears Results in Adoption of Cubs

At the McNeil River where a number of bears assemble to fish, an interesting episode was reported by Erickson and Miller (1963). At this location the assembled bears tend to become accustomed to one another. On one occasion a mother with three spring cubs were assembled at the falls with three other bears. Later, another female with three spring cubs appeared and entered the water a short distance from the first female who was also fishing. The two litters of cubs mingled on shore. The second female caught a salmon and rushed up a bluff and out of view. Soon, the first female approached and inspected the six cubs whose odors had perhaps mingled a little. Anyway, the strange cubs were accepted. The female crossed the river with the six cubs and resumed fishing. Later, the second female joined the six cubs and was attacked by the first female. The fight terminated when the second mother entered the water to rescue a cub that had fled into the river and was being swept away. She returned to the site of the fight and followed the trail of the first mother who had left with the remaining five. The first female was seen with the five cubs for several days. This incident ended quite differently from the one I witnessed in which a mother killed two cubs. However, the circumstances were quite different.

Troyer and Hensel (1962) document several cases of cannibalism by brown bears on the coast of southern Alaska. They conclude, "Apparently cannibalism is more prevalent during the breeding season when males are seeking sows. Large males are usually involved and small or newly born cubs are frequently prey." In the incidents they describe, the victims were usually fed upon. In none of the bear-inflicted deaths I have described was there evidence that killing was for food and carcasses were fed upon but little.

The preceding data on joint occupation of range show that several bears may wander over a common range. The incidents that I observed occurred chiefly on Sable Pass, an especially favored area during the summer season. Over most of the park, bears are spaced more widely than at Sable Pass. My observations show that bears tend to move apart when they are near each other, but that proximity may occasionally cause some mortality. It is said that man's greatest enemy is man—likewise, bear's greatest enemy other than man is bear. There may be a tendency for an automatic, self-regulation of population among grizzly bears.

Movements of Transported Bears

At times, especially if garbage is available, bears become troublesome at camps and other habitations. Repeated association among bears breeds an excess of familiarity. When this occurs, bears are removed to distant areas. Observation of such transplanted bears gives some insight into their homing tendencies and abilities.

Some Bears Remain Where Released

In 1960 two large, chocolate cubs, about 3 years old, were close companions at a garbage dump 6 miles from the McKinley Hotel. They fed daily on the hotel garbage that remained intact after some attempt at burning had been made.

To terminate the dump feeding by these two bears one was trapped on 26 August and released at East Fork River, about 36 miles to the west. When the bear was released from the trap, he stood a moment, walked stiffly across the road, and climbed the slope above. When some distance up the slope he stopped, raised his nose, and gave the air currents a prolonged testing. As he moved higher, he made an occasional random bite at the berry bushes along the way. On the skyline he gazed over the country, contemplating the river bars, the rolling tundra, and the many ridges and peaks, as though appraising his new world, and perhaps wondering where the garbage dump was hidden.

His partner was trapped and released on 14 September, at the same location as the 26 August release. I did not see this release so did not witness his behavior, but I was told that he also climbed the ridge above the road.

On 20 September I saw the two chocolate bears about 2 miles east of where they had been released. The bears had found and recognized each other and had remained in the new area, at least temporarily. They fed on berries and were quite tame.

On 24 September the two bears were on the same ridge where they had been released, and were industriously digging roots about one-half mile from the release point.

On 1 September 1961, I saw two bears together that resembled those mentioned above—one dark chocolate and one light chocolate—about a quarter-mile from the spot where the twins were released in 1960. These bears were seen in the same area on 6 September. They were very tame. On 7 September the twins were seen 2 miles to the east. On 9 September they were about 2 miles to the west of their release point, and on 13 September were again seen in the area. On 20 September they were 2 miles east of their release point, and on 24 September were digging roots, as they had the previous fall, about one-half mile above the release spot.

Thus these two bears remained in the general area where they were released for a period of at least one year.

Some Bears Travel When Released

A young bear, trapped at the Morino Campground near the hotel, and released at Milepost 42 on 4 September 1961, apparently did not remain in the area. On the evening of 5 September, Isabelle Woolcock (photographer and keen observer) saw it dumping the garbage can boldly near her camper at Igloo Campground (Milepost 32). She stated that she knew the bear when it was raiding the Morino Campground. On the same day I had seen two bear scats strung along the road between East Fork (Milepost 42) and Igloo Campground (Milepost 32) that contained remnants of garbage, suggesting that the bear released at East Fork had been hurrying along the road toward the Igloo Campground.

On 24 September 1961, a small bear, about 3 years old, was trapped near the hotel and released 39 miles to the west, on Sable Pass (Milepost 39). When released at 11:00 a.m., he ran far up the slope, then galloped east and north for 3 miles down Igloo Creek. I saw him at intervals along the way as far as Milepost 34. At 3:30 p.m., when I stepped out of Igloo Cabin (Milepost 32), this bear was 7 or 8 yards from the door. He continued moving through the woods, northward. How far the bear continued was not known, but he did not remain where he was released. If he continued following the road in the direction he was going, he would have returned eventually to the point of capture.

In 1969, a relatively small, dark bear that had been visiting the road camp at Toklat was trapped on 21 August and released 19 miles (by road) west. The next day we observed a small, dark bear at Milepost 55 moving eastward and limping badly on one forefoot. Later, we saw it cross the river and move about 2 miles east of the Toklat road camp. We noted a white splotch on the bear's muzzle which later confirmed the identity of this bear as the one transplanted the day before. Apparently, some white paint was spilled on the bear's head by mistake. This bear returned from about 19 miles away within 24 hours after release.

Bears that have been feasting on garbage have a strong incentive, perhaps, to return to their former haunts after they are moved elsewhere. Even so, some may remain in their new locale as did the twin cubs described above.

4
The Family

Mating

Breeding Season

The breeding season of grizzlies in McKinley National Park extends from mid-May to early July. The earliest date on which a breeding pair was seen together is 14 May, and the latest is 10 July. In 17 seasons I have seen 51 pairs. Of these, 35 were seen only once.

I have divided the breeding season into five periods, and tabulated the number of pairs seen in each period. (Nine pairs were seen in two of these periods and two pairs in three of the periods, so these 11 pairs are recorded in more than one.)

Pairs seen in different periods:

14 May to 22 May	3 pairs
23 May to 31 May	10 pairs
1 June to 10 June	19 pairs
11 June to 20 June	16 pairs
21 June to 30 June	15 pairs
1 July to 10 July	4 pairs

Thus a major part of the breeding, according to these records, takes place in the last week of May and during June.

Duration of Matings

The duration of the mating period is variable. A female that bred with two males was attended by one of the males for about 23 days. A male that mated with two females over a minimum period of 21 days was accompanied by one for 15 days (for the last 3 days this female apparently had finished breeding), the other one also for 15 days. Other pairs were seen over a period of about a week or longer but their total breeding period was not ascertained.

Mating among bears seems to be a strictly physical activity. I have seen little indication of fondness between the sexes. When the relatively short mating period is over, no further companionship takes place. The male is not needed as a provider of food—which possibly is a factor in the business-like mating arrangement of grizzlies.

Breeding Interval

My evidence indicates that females followed by cubs have a minimum breeding interval of 3 years, but generally longer. This minimum 3-year breeding interval may be deduced from the following data. Only 3 of the 54 breeding females had cubs nearby. Each of these three sets of cubs was at least 2 and possibly 3 years old. If mothers bred every other year, there would have been yearlings with mated pairs or away from their mothers on their own. However, all yearlings noted were seen with their mothers, and with mothers that were not mated.

Sixty-nine mothers were seen with 2-year-old cubs. Of these, 25 still had their cubs after the breeding season (10 July), and others with cubs were seen near the end of the breeding season. For these 25 females the breeding interval was at least 4 years. Some families with 2-year-old cubs that were last seen early in the summer may well have been with the mothers after the breeding season but escaped observation. Five mothers were each followed by a 3-year-old cub so their breeding interval was known to be at least 4 years.

One mother was known to get rid of her two 2-year-old cubs in May and breed again, so her breeding interval was 3 years.

Breeding Behavior

Males Searching for Females: During the mating season one occasionally may see a big male traveling steadily with long deliberate strides, obviously searching the countryside for a female. Many of the females, those with spring and yearling cubs and at least some with 2-year-old cubs, are not available. As a consequence, each year the males have only a fraction of the female population from which to find a mate. This situation results in considerable searching by the less fortunate males who do not happen to meet up early with a receptive female. A few observations of males seeking mates may serve to show some of this behavior. I expect that a female may occasionally be on the lookout for a male, but more often a male is with the female before she is ready to breed.

On 26 May 1961 at 7:45 a.m., I started to watch a large, dark male striding westward on the flats of Polychrome Pass. He seemed to be following a trail. He appeared on top of a spur ridge and on the way down negotiated a snowfield in two long slides. When again out on the flats, he had his nose to the ground and made many turns as though following a trail. At 9:25 a.m., he arrived at a spot where I had earlier seen a bear that appeared to be a female. It was apparent that he was following this bear's trail, but she had disappeared some distance to the south. Later, he circled a few times, made a short rush toward 2 caribou, then ignored 35 or 40 caribou. At 12:30 p.m., he laid down on an isolated patch of snow just large enough for a bed. At 3 p.m., he began to move again, came to where he had apparently lost the trail earlier, and circled there. By now the female was far to the south, digging roots. I left the scene at 4:30 p.m., leaving the male circling and gradually approaching the other bear. For almost 9 hours the male had not eaten, very unusual under other circumstances, but it appeared that he was concentrating on finding a mate.

On 12 June 1961, a large, dark male hurriedly crossed 2 miles of flat country, made a circle near the north end of East Branch Range, traveled a half-mile toward me, and disappeared in the spruces. In 10 minutes he returned and moved over toward his earlier trail. He seemed eager and in a hurry. He alternately walked rapidly, trotted, or galloped. When trotting, his huge, fat bulk rolled loosely. Most of the time he seemed to be following a trail. He entered a pond, shook himself in the water, but on emerging did not take time to shake, his soaked hair lying flattened against the hide, the water streaming off as he walked. For 1½ hours he hurried along, apparently quite impatient. The day before I had seen a blond bear here which could have been a female, judging from its size and general appearance.

On 27 May 1960, a large, very blond male traveled up Igloo Creek and over Sable Pass in the early afternoon. Later, I saw him moving westward along the base of Sable Mountain. At 8:30 p.m., he was walking steadily along the base of the East Fork sheep hills. He dropped down on the river bar and at 9:40 p.m., still traveling steadily, disappeared around a bend in the river. During the afternoon and evening he had traveled up Igloo Creek, then 4 or 5 miles across to East Fork (I do not know what, if any, detours he may have made) and down East Fork which flows roughly parallel to Igloo Creek. His travels had taken him on a u-shaped course. I saw him take only a few bites of vegetation during the 2 hours that I watched. He was a stranger and I did not see him again.

On the flats of Polychrome Pass, on 31 May 1963, I saw a large, dark male following a small, blond female a half-mile ahead of him. The female turned sharply toward a group of about 100 caribou and chased after them at full gallop. She appeared to have captured a young calf that had not tried to escape, for she stopped and fed. In the meantime, the male followed slowly, occasionally stopping to scratch his back on a boulder. When he was 200 yards away from the female, she stood up, looked at him, and galloped away with the remains of a carcass in her jaws. Some distance up the slope she turned at right angles and was lost in the broken terrain. He followed her trail until he also disappeared at the same spot where I had lost sight of her. Apparently, these animals had not paired off at this time, but the male's persistence suggested that the female soon would welcome his attentions.

One big male seen toward the end of the breeding season apparently had ceased to search. Perhaps he had finished a mating or given up for other reasons. From 25 to 28 June he fed and rested on the carcass of a bull caribou. The 4 days of gorging were perhaps just what he needed after a strenuous season.

Travels of Mated Pairs: Movements of mated bears probably vary a great deal. If a pair mates during a food transition period, it is possible that some shifting of range occurs, such as from a favorite rooting area to another area where green forage is becoming available. On 14 May 1961, a pair was seen to move about 2 miles to a choice rooting bar about 1 mile long, where it stayed for 2 weeks.

A male, mated with two females, remained in an area less than 3 miles in diameter for 21 days. These bears were seen at this location before and after this mating period. Another mating pair was observed to shift its hub of activities a distance of about 10 miles.

Quite often, I observed a pair on the move, the female leading and the male following methodically some distance behind. On 15 June 1948 I saw for the first time a large male and a female on Sable Pass. She was moving away in long gallops as he galloped after her. Once she stopped and faced him as she sat on her haunches; he stopped until she again hurried away. They disappeared over a distant skyline and I did not see them again. The fact that I saw many pairs only once in this open country suggests that their movement is considerable.

On 28 May 1963, as Dr. Frank Darling and I were climbing Primrose Ridge, we saw a pair also climbing the ridge, moving up the draw to one side of us. The female moved along steadily in the lead. We lost sight of them when they went out of view toward the top of the ridge, and we did not see them again.

On 24 June 1963, as I was approaching the Toklat River, a dark female galloped off a ridge and out on the river bar. Some 100 yards to the rear she was followed by a big, dark male. Out on the bar he moved to the up-river side and she turned and followed the stream northward for a few hundred yards, climbed far up a slope, made a loop and returned to the bar. She crossed the river bar, he all the time 100 yards or more behind, and thcy disappeared into a wooded slope. For much of the time spent on the flats, the bears galloped, crossing the streams with much splashing.

Thus we find mated bears both sedentary and traveling. The movements observed probably are somewhat dependent on their mating stage.

Playing Coy: On 17 May 1961 at 8:40 a.m., I discovered a large, dark male with a blond female on the East Fork River bar. They were lying about 30 yards apart. He sat up, looked around, then approached the female. She walked away 10 yards, made a sharp turn, and lay down. He stopped and stood about 8 yards from her—too near, for she walked slowly away. She led the way to the edge of the river bar and fed on roots. He moved about in a draw and behaved as though there was an interesting scent in the air, which there was, for after he had trailed around considerably and dug roots, he found the carcass of a calf moose in the brush. Once, when he passed about 10 yards in front of the female, she made a bluffing lunge forward, striking the ground hard with both forefeet as bears often do when they are warning or bluffing humans. At times they fed 10 yards apart. At 10:25 a.m., both bears moved out on overflow ice covering parts of the bar. He quit following her and moved up river. When he sat down she moved to within 4 or 5 yards, but when he started to move toward her, she jumped away a few yards, then approached him and repeated the maneuver, as though she was teasing him. They again dug roots and moved a quarter-mile up the river bar. At 11:30 a.m., he followed her and she retreated. He moved away to a bluff where he fed on a moose calf. A little later, both bears were out of our view. This male spent considerable time feeding on roots and the carcass of a moose calf.

On 18 May these bears were near the place where they were last seen the previous day. The big male crossed the ice; 5 minutes later the female followed. On shore she walked close to him but then retreated when he moved toward her. When she went out on the ice, he spent 5 or 6 minutes herding her away from the shore. Once she made a romping charge toward him and he made a few jumps away, but then approached her. She finally reached shore and dug roots whenever she had a few minutes of peace. She crossed and re-crossed the wide river bar to avoid him, continually acting coy. We left them as they continued maneuvering in this fashion.

The following day, a quarter-mile up a long slope, the two bears were continuing their love play. When the male stopped following her, she would move toward him. When he lay down, she sat 25 yards from him. Thus we left them and did not see them the following days.

Two Males Breed with One Female: I have described elsewhere (Murie 1961) the behavior of two males that bred with one female. The female first was attended by a small male that bred with her. Later, a large male followed the pair persistently and also was observed breeding with the female. The last day the bears were observed, the large male apparently tolerated the smaller male, a rather unusual situation. The mating period extended from about 19 May to at least 10 June, the last day the bears were seen together.

Males Acquiring More Than One Mate: On 7 June 1967, a large, brown male and a blond female were seen digging roots on the Toklat River bar. He followed the female several times but, for the most part, seemed content to feed or rest nearby. A second blond bear, perhaps slightly smaller than the female, also was digging roots about 300 yards from the pair. I watched these three bears again on 9 June for about 7 hours when all three were again digging roots at Toklat. The second blond bear approached the female several times and they touched noses amicably. The male was not too concerned about the second blond, but when he approached them, the blond would run a short distance away. The male showed more interest in the female, following her repeatedly, and standing over her when she sat down. Eventually, all three bears lay down on a hillside, the male and female side by side, with the other blond about 300 yards away.

When I reached Toklat at 3:15 a.m. the next day, the male and female were copulating and the other blond was about 400 yards from them; he walked about a mile and stopped. The male remained mounted on the female for 25 minutes, occasionally thrusting. They separated and the female began to dig roots. The male moved off out of sight for about 45 minutes, then reappeared on the river bar, moving in the direction taken earlier by the other blond bear. He smelled the ground at intervals, as though following the trail of the blond, and walked at a rapid pace quite different from the usual slow, ponderous gait of males. He soon joined the other blond and they grazed briefly before the blond moved into a draw, followed by the male, where both lay down. After a half-hour rest, the blond led the way as they moved out onto the river bar. These bears were not seen subsequently; perhaps they moved farther up the river bar and onto the slopes and ravines of Divide Mountain. Apparently the second blond was another female. This male concentrated his attention on one of two females, then turned his attention to the second female after consummating his first affair. No overt conflicts were apparent while all three bears were together.

In 1959 I observed a large male, crippled on a front foot, keeping company with two females, one blond and the other dark, with a limp on her hindfoot. The blond female was seen with the male from 20 June to 4 July. On 2 July she was a half-mile from the male and on 4 July, 75 yards away, after which she moved off. The dark female was with the male from 26 June to 10 July. For about a week the male maneuvered with both females. During this period, the females seemed oblivious of one another, the concern of each being the male. It probably was a satisfactory arrangement, for it gave each female some respite from the male's attention. This mating has been described in some detail elsewhere (Murie 1961).

Crippled Male with One Female: The crippled male was not seen during the three summers after 1959 when it mated with two females. However, on 11 June 1963, this dark, crippled male, still limping severely on the left foreleg, was observed following a blond female 3 or 4 miles from the area on Sable Pass where he had rendezvoused with two females in 1959. The pair was moving up the East Fork River bar about a mile above the bridge.

The following day, 12 June, the pair was observed near the East Fork cabin coming down the river close to the east bank, the female in the lead. When the pair came to a stream directly below where we were watching, the female turned and followed the stream eastward.

On 16 June the pair was seen less than a mile up the creek. The male was resting on overflow ice, the female engaged in feeding on roots. He followed a parallel course to where she first fed downstream and then upstream, and lay down on the ice opposite her. A few times he intercepted her, a herding maneuver. They frightened out of the creek bottom a cow moose with a calf but may not have been aware of the presence of these two. Later, the female went up the slope, moved farther east, and returned to the creek. When the male came looking for her, he stood uncertainly a little distance off her trail, uttering some breathy "oofs." He returned to the creek bottom and later the two were together again. In the evening they were in the same area, the female feeding on roots and the male lying on the ice. I had not seen him feed during the day.

On 17 June the pair was seen on the East Fork River bar, and later the bears climbed up among precipitous cliffs, the male resting on a ledge 10 yards above the female. They rested from 6:30 a.m. until 11:00 a.m., at which time the female began to climb, followed by the male. They went over the top of the ridge, dropped down on the far side to the river bar, and returned to the stream where they had been seen the previous day. The male was seen to eat a few bites of green grass. He seemed quite gaunt. The pair was not seen after 17 June (Fig. 20).

Pair Passes Through Caribou Herd: On 6 July 1948, I discovered a pair of bears asleep at the base of a ridge east of Toklat River. Later, the female fed for a half-hour and then alternately galloped and walked across a flat where a large number of caribou were feeding. The male followed. The caribou moved aside to form a narrow lane for the passage of the bears. When the bears were recrossing the lane, the male mounted the female for 40 minutes. She kept swaying from side to side as though attempting to get free. When they separated, they rested for a half-hour and then continued on their way through the caribou herd, seemingly oblivious to their presence.

Herding the Female: Males were seen frequently behaving as though trying to keep a female mate from traveling by circling in front of her whenever she started to move away. Females seemed to exhibit much perversity and coyness in maneuvering.

Fig. 20. The crippled male patiently following his female mate.

On 9 June 1962 at 4 p.m., I discovered a large male with a small female. They were resting 4 or 5 feet apart. After a half-hour, she started across an extensive snowfield. He galloped after her. Reaching bare ground, she alternately fed on berries and new grass and moved away from the male. He herded her, moving in front to head her off, then sitting on his haunches or lying down. Soon she again would begin to move away. He managed to keep her in a small area, so that after an hour she was back where they had been resting when first seen.

At 6 p.m., the female rested and the male sat on his haunches 10 yards in front of her. He kept lifting his muzzle as though catching a scent. For 20 minutes the male sat in front of the female, biding his time, following grizzly ritual. He moved 5 yards to one side and again sat watching her, head held forward. She lay on her stomach, muzzle on ground. At 6:40 p.m., he lay down. At 6:50 p.m., he walked three or four steps toward her. She retreated 50 yards and lay down, and he sat on his haunches. A small, grazing bear appeared 150 yards away; when it saw the two bears it retreated hastily. The female fed again and moved along, and the male continued to head her off. I left them at 8:15 p.m., during this maneuvering.

The following day, when I passed them at 3 a.m., the two bears were in the same area. When I returned at 8:15 a.m., they had moved about 1 mile—the female had made a long run, according to a tourist who had watched them. During the day, the pair behaved as they had on the previous day. On these 2 days I had not seen the male feeding, and the female fed very little on the afternoon of 10 June.

On 11 to 13 June, I did not see the pair but they probably were in the area somewhere. On 14 June they were about 1 mile from where I had seen them 4 days earlier. He was herding her but she stayed about 50 yards away. They both fed on this day. Another large male, a mile away, moved toward them but later angled away and they did not encounter each other.

On 15 June, in the morning, the female followed 50 yards behind the male; they entered a hollow and were not seen during the hour that I waited. At 6:00 p.m., the two bears were feeding 100 yards apart. On 16 June the pair had moved about 1 mile. When discovered, the male was resting on a snowfield and when he started walking westward, she followed him at a distance. Later, I saw him feeding while she rested and rolled on a snowfield. When he started walking toward her, she galloped 50 yards and resumed feeding. On 17 June the bears had moved about a mile, back to their 15 June location. The male was feeding and she was lying on a snowfield 100 yards away. A little later both moved out of my view. On 19 June the two bears were at least 300 yards apart and not in view of each other.

These bears were seen from 9 to 19 June and on the latter date their breeding association appeared to be terminating.

Toklat Mating and Fight: When I arrived at Toklat River on 7 June 1962, I saw a photographer, accompanied by a prospector and his wife, changing film while two male grizzlies were having an altercation about 75 yards or less away. The bears came together on a snowdrift where they raised up on hind legs and the larger bear pushed the other who sprawled backward into the slushy snow, which splattered widely. There was a scuffle and they separated. I did not see any biting. The rush and push of the larger bear overbalanced his opponent. The vanquished bear walked away a few steps and stood, then tentatively began to leave, moving a few steps but lingering for a few minutes. The larger male walked 20 yards or more away, then swung back to the spot of the encounter. The prospector told me that there was much snarling, growling, and foaming at the mouth. The defeated bear moved downriver, below the Toklat River bridge, and disappeared in shrubbery. The prospector said that he saw a pair of bears later, breeding below the bridge, and he assumed that the defeated male was one of the participants.

The victorious male moved out toward two smaller bears that appeared to be females. He came up close to one of them and seemed to nose her, then continued on 200 yards to the other bear and approached to within 3 or 4 yards. After a short pause, he returned to the first female who galloped toward Divide Mountain, the male galloping after her. The female led the way up the west slope of the mountain and then continued southward on a high contour halfway up the slope. The female was usually in the lead but occasionally the male was ahead and at times they traveled parallel, 50 yards or more apart. When the one behind stopped briefly to feed, the other waited or turned back. After moving across the mountain for a mile or more, they descended, fed briefly on the river bar, and soon crossed over the broad bar and fed slowly, moving north. I lost them at the mouth of a small canyon. They had traveled for over an hour and the other female followed the same route, starting out after the pair had left the area. When she came down on the bar she fed for 2 hours, was still feeding when I left, and was alone, about a quarter-mile from the pair.

The following day I saw what appeared to be the same pair feeding on roots on the river bar west of Divide Mountain, the two bears about 100 yards apart. When the male moved toward the female, she started galloping, and he galloped after her. After moving about 200 yards, she stopped to feed, and he stopped to feed about 150 yards from her.

A Second Male Takes Over: On 18 June, 1964 about 8 a.m., on the north slope of Cathedral Mountain, a male was discovered breeding with a dark female. After 12 minutes, they separated and both lay down. An hour later he approached within 5 or 6 yards and she moved away. He followed, keeping below her and preventing her from moving down the slope. She lay down, and a few yards below, he did too.

The following morning I saw the male following the female up the slope. When she stopped, he approached and mounted her. In 4 minutes they separated and faced each other. Both lay down. One half-hour later they moved 50 yards and lay down 10 yards apart. Twenty minutes later the female moved 25 yards and sat down on her haunches. The male approached and stood with head against one of her hips. Both lay down. In a half-hour he was following her, and when she stopped they touched noses and sparred briefly. After a few steps and more sparring with heads, he lay down. She sat on her haunches for 18 minutes, with head between her front legs, nose almost touching the ground, as though dejected. She next moved 20 yards and again they put their heads together and gently sparred or fondled. He tried to get behind her and she kept turning so as to face him. In 15 minutes he lay down and she stood with nose to ground for 25 minutes, at times swaying her head from side to side. At 11 a.m., the male approached the female and they nosed each other for 2 minutes. He lay down and she stood with her head down as before for an hour. At noon she moved 20 yards—he followed and lay down 15 yards below her, and she stood again with lowered head for 15 minutes then lay down. Later they fondled briefly, then he lay down, and after standing 10 minutes with lowered head, she also lay down.

After almost 3 hours, she sat up on her haunches with lowered head; he put an arm around her, moved off a few steps, returned to nose her head and neck; but she would not let him get behind her. Between 4 p.m., and 9 p.m., this maneuvering continued. They nosed each other five times but she continued to keep him from getting behind her. For periods of 15 or 20 minutes, she would stand with nose almost touching the ground. Neither bear had fed all day. I had never seen a female so steadily approachable—apparently she was at the height of her breeding period.

At 5 a.m., on 20 June the pair was near where I had left them the previous evening. While he rested, she moved one-quarter mile down the slope. When he looked around and found her gone, he nosed about until he found her trail and followed, occasionally breaking into a lope. Below the pair was a blond bear, another female. I had seen her on the slope the day before but not near the pair. She raised her nose, apparently getting the scent of the pair, and started walking up the willow slope toward them. She lay down about 200 yards below the male and the brunette female who were maneuvering on the slope. The three bears rested for most of the morning. At 1 p.m. when the male started to walk toward the blond, the brunette took the opportunity, it seemed, to escape, at least temporarily, for she moved away at a fast walk. The male, discovering her hurrying away, galloped after her. She saw him coming and stopped, and he moved to a position on the slope below her. The male then started watching the blond down the slope, and the brunette again walked rapidly away. He sat on his haunches with head down, as though contemplating. He decided to go to the blond resting below him and as he approached she galloped away. He sniffed thoroughly the bed she had left, then galloped after her and they disappeared from view. Soon they were seen far up the slope, he following 25 yards behind on a steep hillside. The blond stopped on an outcrop; he now hurried down the slope to where the dark female had been and followed her trail at a fast trot or walk until he found her. At 4:15 p.m., the pair became hidden in the willows and later the blond also was hidden. At 10 p.m., the male was seen following the blond. She waited, they touched noses, and he lay down while she sat on her haunches. A half-hour later she was still on her haunches and he resting as before. The behavior of the brunette was different from the previous day's; today she was standoffish and kept her distance.

At 5 a.m. on 21 June, far up the slope, the male was herding the blond from below. Fresh tracks in snow patches showed that bears had been higher on the mountain during the night. In the course of the morning the male moved from one female to the other, sometimes seeming uncertain which to attend. He was closest to the dark one most of the time, and she was again more tolerant of his attention. However, while the blond was receiving his attention, she took the opportunity to hurry away.

At noon the brunette and male were lying close together, with the blond lying a short distance up the slope from them. Far down the slope, near Igloo Creek, I saw a large, very dark male climbing toward the three resting bears. About that time a heavy rain and mist hid the bears. When it cleared, the big male was where the three bears had been resting, and the three bears were scattered. The lighter male was on an adjoining ridge, the two females higher on the slope, one to the east and the other west of the big male.

Later, the big male began to walk with a ponderous, slow gait toward the brunette, the smaller male disappeared eastward, and the blond female moved west. The brunette went over a ridge and the big male lay down. After resting 1 hour and 45 minutes, he followed and was soon out of sight.

On the following afternoon the big male and the blond were resting 10 yards apart. Soon the brunette appeared among the willows nearby and the big male joined her and mated with her for 25 minutes. She then moved westward, the male following, and the blond female brought up the rear. The three bears disappeared and were not seen together again.

Male with Family: In 1960, the behavior of a mother with two 2-year old cubs suggested that she had mated with a male and was still on intimate terms with him. Apparently, two females had been involved with one male. This relationship will be described in some detail, although later records of similar behavior may change the interpretation that now seems most plausible.

On 30 June 1960, I saw the family for the first time that year, although I had seen them frequently in the same area the year before. Two hours later I saw a tall, rangy male bear in the distance, moving nearer as he fed. When he was 150 yards upwind from the resting female, she raised her head as though getting his scent, but then relaxed. When the male had fed to within 50 yards (about 2 hours after I first saw him), one cub saw him and galloped away, the other cub following. The mother followed the cubs for 100 yards at a gallop and the male galloped after them. She stopped and faced him, muzzles only 2 or 3 feet apart. They moved so that they were standing parallel, shoulder to shoulder, a few feet apart. Then she walked slowly to her cubs 100 yards up the gentle slope. Again, the male chased and again she faced him. In a few moments she moved about 30 yards, her cubs joined her, and all four bears fed. One cub fed within 10 yards of the male, and later the female was as close to him. A little later the mother seemed to panic and ran a short distance then walked into a green swale 200 yards from the male.

Later in the afternoon, the lone male was only about 20 yards from the family. The mother made a threatening charge toward one of the cubs as though to shoo him away, and a few minutes later charged to the male and they sparred, gently it seemed, with jaws wide open. The mother then turned and fed only 5 or 6 yards away, and the male sat on his haunches, as males often do when they are near a female. The cubs had retreated 100 yards when they heard the altercation. Soon all the bears were relaxed, and grazed.

I watched these bears from about 11 a.m. until 9:30 p.m., when a fog enveloped the area. Twice during the day the cubs were seen nursing. It was obvious from their intimacy and relaxed behavior that the family and the male had been associating for some time.

When I saw the family and male on 1 July, they were from one-half to a mile apart. On 2 July the male was not seen but what appeared to be a blond female fed in the area about a half-mile from the family, and on 3 July only the family was seen. By 4 July the family was a mile or more from the male.

On 5 July the blond female grazed 150 yards from the family. When she started walking toward the family, it moved 50 yards to one side and she continued forward to a distant, favorite green swale, where she fed. The unperturbed behavior of these bears was unusual and indicated a past familiarity. Later in the day the male, who had been feeding a mile away, moved close to the family, which was first startled, then relaxed, and resumed feeding. A few hours later, the three units were spaced about 300 yards apart. I was absent for about 20 minutes and on my return a tourist told me he had seen the male and the blond female "charging each other" with much roaring. These two had moved 200 yards apart, and the family had left.

For the following 3 days the bears were seen but were well-spaced when I saw them. On 9 July the male, moving across the top of Sable Pass, saw the family 200 yards away and galloped toward it. The female stood her ground, but the cubs galloped 200 yards down the slope where they were soon wrestling. The two adults stood about 10 feet apart for about 3 minutes and then the female edged away, joined her cubs, and they walked steadily away another 300 yards. The male began to feed.

At 8 p.m., on 11 July, the male and the blond were feeding about 100 yards apart. The family was resting 300 yards from the blond bear. The brown male fed within 30 yards of the blond bear, started walking toward her, then charged a short distance. The blond stopped feeding and stood for 5 minutes, while the brown male remained in a hollow a few yards away. Later, the bears sparred with open jaws, a foot or two apart. They then relaxed and fed 10 or 15 yards apart for 10 minutes, facing away from each other, seemingly unwatchful and indifferent. They faced each other again 3 or 4 yards apart, then again fed. This was odd behavior for which I cannot account unless the male also had been mated with this blond bear.

On 12 July the family and the brown male fed about 100 yards apart. The cubs played roughly later in the day and galloped a quarter-mile from the mother. She and the brown male both started walking toward the cubs. The female, upon seeing the male advancing, veered toward him and then he walked toward her. When they were 15 yards apart, they both stopped. Then I heard the female utter a prolonged, roaring growl, and soon she moved toward her cubs. The male resumed feeding in a casual manner. The following day, the 13th, the brown male was still near the same place, as he was on the 15th and the 21st. The family was not seen again until 11 August, when it was in the process of breaking up.

It appears that the brown male had mated with two females and that I had observed the waning of the breeding period. If this were so, then a female with 2-year-old cubs had bred while still on good terms with her offspring. Whatever the relationship, this was one of the few times I observed such behavior between a male and a female with cubs.

Summary: A general pattern in the behavior of breeding pairs emerges. Females tend to pay scant attention to males when they encounter them early in the breeding season. Males, on the other hand, are intent on maintaining their association with a female, although the presence of a second female sometimes complicates their actions. At times, a male actively herds a female in an attempt to prevent her from escaping his attentions. Eventually, a female becomes more tolerant of the male. This increased responsiveness seems necessary for breeding to occur and perhaps is related to the onset of sexual receptivity during estrus. The female's initial aloofness may function to retain her availability to other, perhaps more fit males that might encounter her and displace the initial suitor.

Nursing

Observations made over a period of years in McKinley National Park have added considerable new information concerning the nursing habits of grizzlies. Of special interest was the discovery that yearling cubs are nursed routinely and to the same extent as spring cubs. Even more surprising was that 2-year-old cubs also nursed routinely, at least during the spring and early summer.

The time of weaning varies with different families. I have seen a 2-year-old cub nursing in late July, and it would not surprise me if some cubs, especially in one-cub families, were still nursing in September. One wonders why the period of nursing is so prolonged, for even the spring cubs feed extensively during the summer on all the foods enjoyed by bears.

To what extent the information gathered at McKinley National Park is typical of grizzlies in other regions and other environments is not known because very little has been published concerning this phase of grizzly habits. Pearson (in Herrero 1972:81) reported that in Yukon Territory weaning does not take place until cubs are 2½ years old, and the Craigheads' data from Yellowstone Park suggest that weaning occurs at 1½ and 2½ years of age with equal frequency (Herrero 1972:81). On Kodiak Island, cubs usually are weaned as yearlings, but at least some mothers continue to nurse 2-year-old cubs (Herrero 1972:81).

Nursing Posture

The mother bear is equipped with three pairs of teats: two pairs pectorally on the chest, and one pair inguinally between the hind legs. Cubs nurse at both pectoral and inguinal teats, but pectoral nursing predominates. Cubs, especially single ones, frequently switch their attention from one side to the other several times during a nursing, and shift from pectoral to inguinal locations. The usual pattern I observed was pectoral nursing initially, followed by a short period of inguinal nursing near the end of the session. When the mother decides to favor the cubs with a nursing, she generally sits down on her haunches and rolls over on her back. While the cubs nurse fore and aft, she bows her neck so that it is raised off the ground and held with muzzle pointed toward the cubs as though to supervise or admire her growing offspring. For humans, the bowed neck would be a strain, but apparently mother bears do not mind for the position is sometimes maintained after the cubs have finished. Occasionally, a mother may relax toward the end of a session and rest her head on the tundra. One mother seemed weary from the start and lay relaxed during an entire nursing. While lying on her back, the mother's legs may be in various positions. The hind legs may be extended somewhat or bent so that knees are raised on either side of a cub nursing inguinally. The arms usually are relaxed along the sides of her body, but occasionally an arm extends over a nursing cub. One female I observed nursing a yearling formed a basket by grasping hind legs with paws (Fig. 21).

A cub may begin to nurse while the mother is in a sitting position but usually she will roll over immediately onto her back. Only a few times have I seen a mother remain sitting on her haunches during an entire nursing period. Occasionally, a mother was observed sitting on her haunches while her yearling nursed a minute or so, and then tip back into the usual posture. Nursing inguinally would seem to be more convenient for the cub when the mother is on her back. On a few occasions a mother nursed the cubs while lying on her side, and sometimes, after she had rolled over on her side from the back position to terminate a session, a cub might continue trying to nurse.

One mother was just too sleepy to stay awake during a nursing. On 4 August 1961 this mother returned to an old caribou carcass of which

Fig. 21. A mother nursing her spring cub, in the typical nursing posture for grizzlies.

little remained except bones. She managed to extract some long, tough pieces of sinew from the leg bones. Her two spring cubs chewed at the bones for a time, but soon wearied and waited for the mother to complete her salvage operations. When, at long last, she turned away from the bones, the cubs set up a loud squalling—they had been patient long enough. After walking three or four steps, the mother rolled over onto her back into the usual posture with head raised and muzzle pointed toward her breast. After a minute, I noticed her head falling slowly to one side, lower and lower. Obviously, she had fallen asleep. Then with a jerk she awoke and raised her head, but immediately fell asleep again. I once saw a mother Rock Ptarmigan falling asleep in this manner while standing in the noonday sun, her chicks squatting nearby in the shallow shade of a miniature bank, her head falling to one side or backward as she slept, and waking with a jerk of her head, only to fall asleep again at once. This napping continued for several minutes just as it did with the sleepy bear. The bear continued to nap to the end of the nursing, which lasted 7 minutes, one of the longer nursings I recorded. The mother terminated the nursing by rolling over on her side. For 15 minutes she slept without a stir, then raised her head for a second and lay back at once for more sleep. The two cubs also slept. It was a warm sleepy day.

Start of Nursing Session

The initiative for a nursing session may be taken by the cubs or the mother but, of course, the mother has the final word. If the mother's physiology suggests nursing, she may inadvertently communicate her readiness to the cubs. For instance, if she stops feeding and stands inert, or starts walking, or sits, or lies down, the cubs may interpret her behavior as an invitation to nurse and come hurrying. These cues are of such a general nature that they sometimes may be misleading. I expect that occasionally even when the mother has not planned on a nursing, the impatient cubs may provide the stimulus for one.

If the mother is standing still, the cubs may wait expectantly beside or in front of her. A cub may poke his nose against her chest whether she is standing, sitting, or lying on her stomach. If the response is slow, cubs of any age may cry and bawl. If the mother is walking, they sometimes keep up a complaining cry as they follow. If the mother is ready to nurse, she may roll over onto her back where she is, or, more likely, walk a short distance first, as though choosing a comfortable patch of tundra for a bed.

Sometimes the mother is quite positive in suggesting a nursing. Once a mother walked to her two resting yearlings, touched noses with each one, walked a few steps, the cubs following, and lay down in a nursing position. This is rare; a mother seldom has time to assume a nursing position before the cubs arrive.

On one occasion play intervened and prevented the cubs from responding to an obvious nursing invitation. When a mother and her two yearlings walked out on a snowbank, the cubs started wrestling and sliding immediately. Over to one side, the mother rolled into a nursing position, head held up as though posing for a picture, but the cubs were having such a good time they failed to notice the invitation. In a minute or two the mother stood up, moved off the snow, and started grazing. The cubs did not know what they had missed.

Some mothers appear to seek vantage points for nursing. In June 1967 I watched a female with two 2-year-olds for several days on the lower slopes of Divide Mountain. At each of the five nursing sessions I saw, the mother climbed to a knobby outcrop before allowing the cubs to nurse.

There are endless little variations in the behavior of the family. On 30 May 1964 I spent the morning on the lower slopes of Cathedral Mountain observing a cow moose chasing away her yearling whose place in her affections had been taken by a newborn calf. While the moose were quiescent, I watched and photographed a pair of Willow Ptarmigan feeding and courting in the tundra. A little after 11 o'clock I discovered a blond mother bear with her two blondish 2-year-olds. They had wandered onto a nearby slope and were occupied with root digging. In half an hour the mother lay down on her stomach between the two cubs that were feeding about a dozen yards from her. The larger cub, always the hungrier in this family, came and poked his nose under her chest. There was no delay, the mother stood up and rolled over on her back and the cub lay between her hind legs concentrating on inguinal teats. The smaller cub, who oddly never seemed very anxious to nurse, continued to feed on roots and delayed for 2 minutes before joining and attacking the breasts from one side. After obliging for 5 minutes, the mother rolled over and stood up. The small cub returned to his digging which he had abandoned so reluctantly for the nursing; the other cub and the mother, after standing for a couple of minutes, lay down. In 10 minutes the mother resumed her rooting but the large cub remained resting. About 3½ hours later, the mother walked to a snowfield, ate some snow, and lay down a short distance from the snowfield. In a few minutes the larger cub walked to the mother and she stood up and resumed rooting. Twenty minutes later, she again lay down and when the big cub approached, she rose and resumed feeding. She apparently fed in order not to be bothered. Ten minutes later she and the big cub were resting 50 yards apart. After another 10 minutes, the mother stood up, walked to the resting cub (the smaller one was still rooting steadily) touched noses with him, ate a little snow, and started climbing the slope. The large cub followed closely for a short distance then stopped and watched her move 100 yards up the slope and lie on a ledge. The big cub then followed and when he reached the ledge, the mother rolled over and the cub nursed for 5½ minutes. The small cub missed the nursing and was still digging by himself 150 yards down the slope when I left the scene 10 minutes later.

The family spent the night on the outcrop. They were still there the following day at 4:30 a.m. A little later, they left the outcrop and started digging roots down the slope. Three hours later, the large cub walked to the mother, she moved away, and the cub walked beside her trying to promote a nursing. The small cub also started following and after they walked 75 yards, they all resumed digging roots. The expectations of the cubs did not materialize. Fifteen minutes later, the mother lay down but walked away when the large cub approached. She lay down again, and the large cub joined her pushing his muzzle under her side and chest. When there was no response, he stood there bawling. Ten minutes later the smaller cub also approached the mother. It seemed that because she was bothered, she started digging roots and a little later walked away. The small cub wasted no more time waiting for a nursing and turned to feeding on roots. The large cub sat glumly on his haunches eyeing his mother. She soon walked steadily across the slope on a contour. Both cubs followed. She stopped two or three times as though searching for a comfortable spot for a nursing and finally rolled over on her back and 5-minute nursing took place. This was more than 4 hours after their departure from their night beds. The cubs had wanted to nurse long before the mother obliged. This mother gave the cubs several false cues and seemed strong willed, resisting being forced into a nursing.

But when cubs take the initiative they sometimes succeed in promoting a nursing. One day in late May a mother and her two yearlings were feeding on roots. After a time, I saw one cub quit digging, walk to the mother, and stand close to her side while she continued digging. It was obvious to me what the cub had in mind and it no doubt also was clear to the mother. After 5 minutes, she stopped feeding and stood immobile. The second cub, perceiving progress, stopped his digging and also approached the mother. A few moments after his arrival she rolled over on her back, and a 4-minute nursing took place.

Displacement Nursing Activity

Occasionally, a mother will nurse her cubs during a period of nervous tension caused by the proximity of another bear.

On 25 June 1962 a mother and two yearlings were feeding on a gentle slope on Sable Pass. A hundred yards below them was a young bear, and 200 yards to one side was another, both bears in plain sight. A third young bear was out of sight in the hummocks 300 yards above the family. After a time, the young bear below the family started walking toward the other lone bear. The mother, seeing the bear moving below her seemed to take fright, for she started walking away rapidly. But after traveling 75 yards, she stopped, nursed her yearlings, seemed to regain her composure, and resumed feeding.

On 13 September 1964 a mother and her yearling were feeding on berries. A lone bear, which I judged to be about the size of the mother or slightly larger, came into view about 150 yards up the slope. He was moving along a contour at a steady walk. After a time, the mother either saw him or caught his scent for she suddenly stopped feeding and galloped into a ravine and to the top of the other side. From her new position she could see the lone bear. She watched a minute or two then rolled over on her side and the cub nursed. After 2 minutes of nursing, the female sat up to watch the bear, who by then had seen the family and was hurrying away, sometimes breaking into a lope. Here again the nursing had taken place amid some excitement.

On 2 June 1965 I saw a mother with her 2-year-old cub come over a rise about 100 yards from a young bear, perhaps a 4-year-old. The bears discovered each other about the same time and stood watching. In a minute or so the lone bear came forward 15 yards and stood up on hind legs for a better view as the family walked a few yards to the top of a knoll, where the mother nursed the cub. The lone bear moved 50 yards away and stood watching for 10 minutes then loped away, later settling to a steady walk. No strong reactions were evident on either side. Neither presented a threat to the other, but the nursing at this time of somewhat heightened tension suggested it was a displacement activity.

Female Hurt During Nursing

On a few occasions a female behaved as though she had been hurt by a nursing cub. The reaction usually was a quick movement and sometimes a growl. One mother just starting to nurse in a sitting position made a quick grab at a yearling's head as though she had felt a sudden pain. Another mother nursing yearlings uttered a sharp growl suggesting she had been ruffled by a cub nursing too vigorously. A mother nursing two yearlings stood up and growled her resentment. A mother nursing a 2-year-old cub held arms outstretched tensely as though hurt, rolled over, and terminated the nursing soon after it had begun. Once, when the mother of two yearlings sat down on her haunches, one of the yearlings pushed his muzzle between her forelegs apparently hurting her for she pushed the cub aside with her paw. The cub stood bawling loudly, not, I expect, so much because of the pain, but because of frustration. In a few minutes the mother walked a dozen steps, an impatient cub on each side, and rolled back into a nursing position. It may be significant and is logical that cubs that hurt their mother were all yearlings or 2-year-olds rather than spring cubs. But any noticeable hurt is rare in any nursing.

Cubs Failing to Attend Nursing

Although cubs are always ready to nurse, there were occasions when one cub in a family of two or three failed to attend a nursing. Two such instances were noted earlier in this section. A nursing usually is enjoyed by the cubs too much for them to miss one without good reason.

In early June one of two 2-year-old cubs followed the mother closely until she rolled over on her back for a nursing. The other cub, who apparently had his jaws stuck together with porcupine quills, was a short distance away chomping in distress and trying to feed on crowberries. He was too bothered by the quills to pay attention to anything else, although later in the day he did take part in a nursing.

I have described elsewhere an incident in which a mother nursed a 2-year-old cub after she and the cub had finished their portions of a caribou calf. The second cub, 20 yards to one side, did not come to the nursing but continued feeding on his piece of the carcass. He may have been too intent on his feeding to note the nursing in progress, or perhaps the taste of meat was a greater attraction.

One year, at the beginning of the berry season, a mother and two yearlings were foraging for berries, each moving about on its own looking for choice bushes. After a time, one of the cubs followed the mother closely and succeeded in promoting a nursing. The other cub, 100 yards away, was so intent on his foraging that he failed to note the event.

Termination of Nursing Session

A nursing usually is terminated by the mother when she rolls over on her stomach. A cub may persist in nursing if the mother rolls over on her side and then she must roll the rest of the way to dislodge him fully. One of three spring cubs once continued nursing after the mother had rolled over on her side and was still nursing when she stood up. The mother then swung her head between her forelegs to dislodge the nursing cub. On another occasion, when a yearling objected to a termination, he grabbed the skin on the side of his mother's head with his jaws and shook it vigorously. One mother while nursing a spring cub jerked as though hurt slightly, immediately grasped the cub by the nape of the neck in her jaws, and lifted him off to the side.

Several times cubs were observed to discontinue nursing before the mother turned onto her side. After three spring cubs had nursed for 4 minutes, one cub returned to a spot 40 yards away where he had been digging roots before the dinner bell rang. The other two cubs nursed an additional minute before the mother turned over. On another occasion one of two spring cubs stopped nursing before the mother called a halt. Yearlings and 2-year-olds also stopped nursing occasionally before the mother rolled over on her stomach. Apparently, they had nursed the mother dry.

Several times the mother retained her nursing posture after the cubs quit nursing. One mother, after her lone yearling had nursed 4½ minutes and departed, held her rigid nursing position 3 minutes more before rolling over on her side. On 10 June 1959 two large 2-year-old cubs nursed 4 minutes and stopped, and the mother continued in nursing position for 2 minutes more. On 30 June 1960 a 2-year-old cub after nursing 3½ minutes, moved off to feed on green vegetation. The other cub stopped one-half minute later. The mother remained in stiff nursing pose for 6 minutes longer as though in a trance, then lay relaxed on her back, with head fallen back on the ground, for 10 minutes. On 4 July this same family was observed again. One cub left after 3 minutes of nursing and the other after 4 minutes. The mother held her nursing position for 2 minutes and then let her head fall and continued resting on her back. Other similar nursings were noted in which the cubs, after apparently taking all available milk, stopped and left the mother holding her nursing position. But usually the mother terminated the nursings by rolling over on her stomach. It appears that 2-year-old cubs are more likely than younger cubs to stop a nursing session of their own accord.

Length of Nursing Session

Spring Cubs: I made nursing observations on 15 females who were followed by spring cubs. A total of 51 nursings were observed, of which

34 were timed. These nursing sessions among spring cubs lasted from 2 to 7 minutes, an average of 4 minutes and 15 seconds.

Yearling Cubs: After noting a yearling nursing in 1953, I spent considerable time determining to what extent yearlings nursed. When it became obvious that all yearlings nursed. I devoted less time to observing yearlings, but data accumulated incidentally. Twenty-five yearling families were observed nursing and 65 nursing sessions were seen. I timed 37 of these which ranged from 3 to 6½ minutes, averaging 4 minutes and 20 seconds.

Two-year-old Cubs: When I learned that 2-year-old cubs nursed, a special effort was made to make nursing observations of these older bears. I was able to check on 32 families in which the cubs were 2 years old. In each case the cubs were seen nursing. The latest date noted for the nursing of 2-year-olds was 26 July but families seen after this date might have been nursing. When the break from the family occurs early in the season due to breeding, nursing probably continues as long as the cubs are with the mother. In one family four nursings were noted on 18 May, and 2 days later the mother and her two 2-year-old cubs had separated.

Eighty-six nursings by 2-year-olds were observed and 57 among 25 families were timed. The shortest nursing session was 2 minutes; the longest, a very unusual one, was 12 minutes; the average was 4 minutes and 40 seconds. The short 2-minute session was terminated by the mother because of the roughness of the cub. As in the other age groups, the length of most nursings was near the average. Forty-four of the 57 nursings ranged between 4 and 5 minutes. Sixteen of the families that were timed had two cubs and nine had one cub. There was no significant difference in nursing duration between one-cub and two-cub families except for one one-cub family whose nursings were especially prolonged in a few instances. I shall describe these for their general interest.

On 22 June 1965 a family broke all records for prolonged nursing. The single cub and mother had been feeding during the day on the remains of an old ram in a short draw near the Toklat River. The carcass had been covered with sod and debris and was partially uncovered when the bears fed. At one point the mother stopped feeding and flopped down on her stomach on the mound. Soon, the cub also stopped feeding on the carcass and pushed his muzzle against the mother's chest. She stood up, walked three or four steps, and lay on her back against a bank at a 45-degree angle. The cub nursed for 12 minutes, shifting a dozen times from one side to the other; nursing was pectoral. The cub, after a brief pause, tried nursing at short intervals for the next 2 minutes. All this time the mother lay relaxed, much of the time with her head resting on the ground rather than raised in the usual position. After lying thus for 22 minutes, she rolled over lazily onto her side.

On 19 June this family had walked a quarter-mile from a carcass and at 7 p.m. the cub nursed for 9 minutes 15 seconds. The cub nursed only pectorally and shifted its operation many times from one side to the other. On another occasion (8 July) this cub nursed 7 minutes 5 seconds. I saw this cub nurse 16 times. The three prolonged nursings made the average period for each nursing 5 minutes and 40 seconds, about a minute longer than the average period for all 2-year-olds.

Nursing Interval

The interval between the beginnings of two or more consecutive nursings varies considerably in all age groups. The average length of interval is shortest in the spring cubs, longest in the 2-year-olds. On the other hand, the length of the nursing session is shortest in spring cubs and longest in 2-year-olds. There appears to be a correlation between length of interval and length of nursing session, that is, the longer the interval, the longer the nursing session.

Spring Cubs: On 23 June 1950 I arrived at Sable Pass at 7:30 a.m., and until late in the afternoon watched a family consisting of a mother and three spring cubs. Five consecutive nursings were observed. The first nursing took place at 8:23 a.m., the mother taking the usual posture. After the cubs had nursed for 3 minutes, the mother rolled over on one side. One cub persisted in his nursing and had to be dislodged after the mother stood up. The other two cubs scuffled, and when one cried as though hurt, the mother made a sudden turn toward them, as though ready to protect them. At 10:05 a.m., the cubs moved close to the mother as though expecting to nurse, but she continued to graze. At 10:50 a.m., the mother lay on her stomach, then rolled over on her back. The cubs nursed for 4 minutes after which she rolled over on her stomach, the cubs resting beside her. In 36 minutes the mother resumed grazing, but the cubs rested for another 20 minutes before they started grazing.

Shortly before 1:36 p.m., the cubs again walked expectantly toward their grazing mother. She walked 10 yards, sat down, and rolled over backward. After the cubs had nursed for 3½ minutes, she rolled over on her stomach. Five minutes later she walked a few steps and nursed the cubs for another 4 minutes. In a minute or two she sat up and started grazing. At 2:05 p.m., the family lay down and rested for 50 minutes. The mother then resumed grazing and the cubs picked a little at the vegetation.

Shortly before 4:20 p.m., the mother walked about 200 yards and the cubs chased after her, expecting a nursing. When she stopped, they crowded in close. She sat down, rolled over, and after a 3-minute nursing she stood up and grazed.

The spring cubs had fed five times at intervals of 2 hours 27 minutes, 2 hours 46 minutes, 9 minutes, and 2 hours 35 minutes. The duration of each of the five nursings was roughly 4 minutes, 4 minutes, 3½ minutes, 4 minutes, and 3 minutes.

The interval between nursings in spring cubs varied from 9 minutes to 3 hours and 34 minutes. Omitting the unusually short interval of 9 minutes, the average length of the 18 intervals recorded for spring cubs was 2 hours and 30 seconds.

Yearlings: The interval between nursings of yearlings was timed 14 times. The shortest interval was 1 hour and 10 minutes, the longest 5 hours, an unusually long interval for yearlings. The average interval was 2 hours and 37 minutes.

On 19 June 1959 I spent several hours watching a mother and her two yearlings, and noted five consecutive nursings. The first took place at 9:15 a.m. The intervals between the five nursings were as follows: 1 hour 15 minutes, 3 hours 5 minutes, 1 hour 10 minutes, 2 hours 34 minutes. The first three nursings were terminated by the cubs, and the mother remained in nursing position for 2 or 3 minutes after they stopped. Twice, a cub started the nursing while the mother sat on her haunches but both times she rolled over on her back. From the time of the first nursing at 9:15 a.m. until 3:30 p.m., the mother and cubs rested. This long rest period during the day was unusual and may have been due to the relatively warm weather, one of the warmest days of the summer. The warm weather also may have caused the mother to be lackadaisical about terminating the nursings for, as noted, she remained in a nursing position after three of the nursings had been terminated by the cubs, an unusual procedure for other mothers.

Two-year-old Cubs: For 2-year-old cubs, 24 intervals between two consecutive nursings were recorded. The longest interval between nursings was 9 hours and 20 minutes, the shortest, 40 minutes. The average of the 24 timed intervals was 3 hours and 33 minutes. The average was raised by two especially long intervals, one of 9 hours and another of over 7 hours.

On 23 June 1965 I observed a mother nurse her 2-year-old cub four times. The first nursing took place about 9:30 a.m. I did not see the start but timed the last 3 minutes. The cub stopped of its own accord and the mother rolled over, walked a few steps, and lay on her back. In 5 minutes the cub approached and nursed for 2 minutes before the mother rolled over on her side to terminate it. This was not a regular nursing. It seemed that the cub had taken advantage of the mother's resting position and she was too indifferent to roll away at once. Not including this nursing, the intervals between the four nursings were: 3 hours 23 minutes; 3 hours 59 minutes; and 40 minutes. The 40-minute interval was much shorter than usual, and the nursing may have been due to a displacement activity. The family had left a ram carcass of which little remained worth salvaging and had fed on vegetation in the willow brush. Later, in climbing a slope

100 yards from the carcass, the mother stopped to watch another mother with a yearling that had moved to the carcass. She lay down on her stomach to watch, whereupon the cub nuzzled at the mother's chest region and she rolled over and a 4-minute nursing ensued. This appeared to be some sort of displacement activity, taking place when the mother was under nervous tension due to the proximity of the other family. Being occupied with the other family, she responded automatically to the cub's suggestion. This female had a generous supply of milk; the previous day the cub had nursed for 12 minutes in one session.

Summary of Nursing Sessions and Intervals

The average lengths of nursing sessions and intervals for the three age groups are shown in Table 6. According to the figures, the average nursing interval is shortest in spring cubs and longest for 2-year-olds. The length of nursing sessions is similar in the three age groups, but somewhat less in spring cubs.

Table 6. Average length of nursing and nursing interval for families with cubs of different ages.

Age	No. of families	No. of nursings	Mean length of nursing	No. of intervals	Mean time between nursings
Spring cubs	15	34	4 min 15 sec	18	2 hr 30 sec
Yearlings	25	37	4 min 20 sec	14	2 hr 37 min
Two-year-olds	25	57	4 min 40 sec	24	3 hr 33 min

Aspects of Family Life

Play

Cubs, before and after leaving their mother, spend much time in play. When there is more than one cub in a family much of their play is wrestling and chasing each other. A mother will spend considerable time playing with a lone cub, especially a spring cub. But twins and triplets are so independent and self-sufficient in their play that the mother is seldom called upon to participate. A lone cub will seek its mother for prolonged sessions of play which consist chiefly of tugging and grasping at the mother's head and neck while she paws gently at the cub.

Young bears cannot resist playing when they come to a late spring or summer snowfield. Even a mother may become frolicsome in such circumstances. (A snowfield affects the lambs of Dall sheep and the caribou calves similarly, and I have seen older animals buck and jump when they came to a snow patch.) I saw one 3- or 4-year-old lone cub run to the edge of a steep snow slope and turn a somersault sliding down the almost perpendicular slope. Another lone, young bear gave a vigorous exhibit of jumping and rolling, and, holding his paws straight out ahead of him, pushed himself down the slope with hind legs. At the edge of the snow he encountered a Willow Ptarmigan which immediately put on a wounded act, fluttering and flying ahead of the loping bear just far enough to keep out of his reach. Isabelle Woolcock watched cubs pushing snow to form balls with which they played, pushing the snowball down the slope of the snowfield and pouncing on it. I have seen similar play and also have seen bears break off pieces of icy snow on the lower edge of an old drift and use them for play, pouncing on and mauling them. Once, two cubs were too engrossed in this kind of play to notice a mother's invitation to nurse. Several times I have seen lone bears roll and slide on steep snowslides, and in the spring one often sees trails on steep slopes where the bears have made long slides, no doubt enjoying the sport.

One year in early summer (June 1963), a mother and 2-year-old cub moved out on a snowfield. They broke into a short gallop, then wrestled and played and rolled, sometimes the female and sometimes the big cub on top. They continued playing after leaving the snow patch, both standing erect on hind legs. Twice the mother made short gallops and then faced the cub for more wrestling. Finally, the cub ran ahead 100 yards, then returned to the female who was feeding, and both fed. This mother had been unusually playful, a mood apparently induced by the snowfield.

Another female and 2-year-old cub indulged in play frequently. One day in June they encountered a snowfield on their wanderings. The cub faced the female as she plowed through the snow pushing him aside. Each time the cub was displaced, he recovered and again faced his mother. This continued until the 40 or 50 yards of snow were traversed. Apparently, the female was not moved to play by either the snow or her cub, but later, after a nursing session, she chased and wrestled with the cub in heavy rain for about 10 minutes; a lone caribou was an interested bystander.

After separating permanently from the mother, twins may devote much time to play. I have the impression that some bears play more than others. Two 2-year-olds that I saw several times during one summer seemed to be in a prolonged wrestling match each time. One day a big bear loomed up on the skyline as they were in one of their bouts. They hurried away, galloping in a large semicircle a mile or more in extent until they gained a prominence far above the big bear. Here they lay watching the bear below them. After an hour, they started traveling and feeding, intermixed with frequent wrestling play.

In families in which one cub is considerably larger than the other, the persistence and roughness of the larger cub may become irksome to the smaller one who has to keep breaking away and running to escape. Sometimes when one of a set of twins or triplets is more pugnacious, the mother will intercede and break up fights between cubs.

On 26 June 1956 I spent some time watching a mother and her large 2-year-old as they fed on Sable Pass. The cub, lagging behind the mother, started a prolonged activity which suggested that he was excessively energetic. He began his exuberant play by moving off 100 yards and rolling over on his back, waving and kicking all four black paws in the air. Because the legs are heavy and the hair long, the feet seemed small, too neat and tiny for the bulk they supported. A willow was uprooted, and as he lay on his back, he juggled it in every conceivable way, holding it with various foot combinations, with front paws, with hindfeet, or with a front- and hindfoot. Using both teeth and claws, he removed some of the bark. Sometimes the branch was held with front paws high above his head as though contemplating it. There seemed to be a search for variety as he played with the willow for about 15 minutes. After this prolonged playing, he grasped the willow in his mouth, shook it vigorously and, still grasping it, pushed his head hard against hummocks, apparently to create some antagonistic resistance to his action. His head was jerked against the ground and he jumped in circles and struck the ground fiercely with both forepaws. As he galloped forward to overtake his mother he pounced in a puddle of water with a splash. He jumped into two more puddles sometimes wriggling and jerking his head. When the mother was overtaken, he grasped her hindquarters. As she rolled over to play, he grasped her by the throat and tried to shake her. Later, the mother grasped the skin around the cub's ears and held the cub as though trying to restrain him. Then they sparred briefly as both stood on hind legs. The mother wearied and started walking. When the cub tried to spar, he was pushed aside with a paw. A second time she pushed him away, and then a third time more roughly, causing the cub to stand disconsolately. The last spank he apparently understood and followed the mother with sober step into a swale where both fed.

One can often see cubs in prolonged wrestling, but I have not seen a cub behave so vigorously, so long by himself as this one. A lone cub, especially a spring cub, does not get enough play from his mother, so he dashes about and jumps in the air with a wriggle, starving for a satisfactory outlet. When tall willow brush is encountered, the cubs often climb among the limbs or lie down and spar with the overhanging branches.

Imitation

Cubs often imitate their mothers even though they may not be learning anything they would not eventually learn on their own. Once I watched a mother run into a clump of willows and roll over on her back in the

Figs. 22,23. This female used a telephone pole for backscratching. Minutes later, when the female had finished rubbing, her tiny spring cub awkwardly stood and rubbed its back on the pole.

middle of the clump. When she moved on and the spring cub arrived at the clump, he also rolled into the willows and lay on his back, with feet pawing the air. When the mother dug herself a level bed on a gravel slope, the cub dug himself a bed at the same spot. I have frequently watched a cub rub its back on a pole after the mother had finished rubbing, or dig around in the dirt while the mother was digging for roots or a ground squirrel (Figs. 22, 23).

In 1969 I watched a female encounter a fallen telephone pole. She looked back at her spring cub, then lay down and rubbed her back, head, and neck on the pole. Her cub approached, watched, and began to rub; he seemed to have no special itches but went through the same motions as the mother, continuing even after the female had stopped and walked a few yards away. A few days later I saw two spring cubs watch their mother rub her back on some hummocks. As soon as she stopped, both cubs went to the hummock and began rubbing in the same fashion.

Family Travel

Usually, the mother dictates the direction and speed of travel of a grizzly family. Sometimes, however, the cubs play a role. A female and three yearlings were seen at Thorofare River on 31 May 1959. The three cubs frolicked and galloped 300 or 400 yards ahead of the mother.

On 30 August a female with three yearlings was seen at the base of Mount Eielson about 2 miles from where this female and three yearlings were seen in May. These cubs were as dark as those in May had been. Three or 4 inches of snow lay on the ground. The mother was digging ground squirrels as the cubs huddled about 100 yards from her, hidden by a growth of willows. She stopped digging, looked around, and dashed toward her cubs. She sniffed them and returned to her digging. The cubs then galloped westward across the high bench. When the cubs were a quarter mile from the female, she saw them and galloped for almost a mile in pursuit. She caught up at a prospector's cabin where the two leading cubs had stopped to investigate. Again the cubs galloped forward, leaving the mother far behind, digging. She again galloped after them until I lost sight of them all in the rough country. The cubs thus influenced the course of travel.

Mother Concern for Her Cubs

Many females with cubs evidence a motherly concern for their offspring. When a mother moves out of sight of her cubs in the course of her feeding activities, she usually returns to make sure that the cubs do not lose contact with her. Sometimes, a mother will cuff a repeatedly laggard cub as though punishing it. When cubs move out of sight of their mothers in the course of their play or feeding, she soon maneuvers to keep them in sight, even moving from one napping spot to another from which the cubs are visible. Cubs that are left behind temporarily or lost, especially spring cubs, may emit a hoarse crying or bawling sound which seems to alert the mother to their plight if she is within earshot.

An incident I observed on 7 September 1964 is an example of a grizzly mother's concern for her cub. A mother and her yearling had crossed the Toklat River bar and climbed a long green slope to the edge of a precipitous rocky dropoff above the road. When I saw the mother again, she was picking her way down among the cliffs. By the time she reached the road, the cub had stopped at a perpendicular dropoff which the mother had managed to negotiate. He maneuvered about for a few minutes, afraid to proceed, then climbed upward, soon disappearing. The mother obviously was agitated. My car blocked her progress in one direction, but after some turning about and uncertainty she jumped to the river bar from a 7-foot steel dike. She was uttering deep, throaty growls as she crossed the river and then the road 150 yards to the other side of me and climbed out of view to find her cub. By the time she had reached the top of the ridge, the yearling had managed to reach the road by another route. He hesitated, jumped off the dike, followed the mother's trail across the river channels, and climbed the cliffs, following her trail. When the mother descended the cliffs a second time and found no cub, she was much concerned. She climbed almost to the cliff where the cub had stopped, returned to the road, and seemed ready to do battle with anyone. She finally dropped down to the river bar, and about that time the cub arrived on the road for the second time. He was afraid to jump off the dike again, so moved down the road and reached the bar below the bridge where the mother joined him. She led the way a half-mile toward Divide Mountain before they started feeding on roots.

A few mothers showed rare lack of concern for cubs. One female with two spring cubs was seen several times at Stony Creek in 1969. She spent much of her time grazing and chasing ground squirrels out of sight of her two cubs, and neither she nor the cubs ever seemed anxious over even prolonged separations.

Despite a general concern for their cubs, females capturing ground squirrels or discovering some tasty carrion seem more eager to satisfy their own appetite than their cubs'. But occasionally a mother does extend her concern for cubs to sharing meat with them.

On 29 May 1965, on the south slope of Sable Mountain, I discovered a dark mother with two yearlings. The companionable cubs lagged far behind the mother, at times 200 yards or more, without attempting to keep in contact with her. They fed on the crowberries that remained on the tiny twigs through the winter. The mother disappeared behind a rise, and the cubs, grazing along on the snow-free patches, angled downward. Soon I saw some caribou appear on a high slope a little beyond where the mother bear had disappeared; it was apparent that the caribou were moving away from her. In a few minutes she also appeared, galloping down the slope carrying part of a caribou calf. At first, I wondered how she happened to come directly toward the cubs, who had moved forward and disappeared half mile away from where she last saw them. But on noting the wind direction, it seemed certain that she was following their scent that was being carried up the slope. When in sight of the cubs, she dropped her load and continued forward 150 yards or more. Then, followed closely by them, she returned to the food item she had dropped. The cubs tugged at the calf remains while she rested a few steps away. She must have been surfeited, otherwise she would have been active in getting her share. When she left the cubs, she probably went up the slope to retrieve remains of a calf killed earlier, one on which they probably all had fed previously.

The family was not seen the following day but was discovered at 7 a.m. on 31 May slightly west of where they had been seen on the 29th. The mother was moving forward again far ahead of the two cubs who had stopped to dig roots. The mother, walking rapidly and purposefully, climbed a side ridge and disappeared behind a cone on the lower slopes of Sable Mountain. Soon after she reappeared from behind the cone, she started loping toward 14 caribou cows that were hurrying away in the distance. Soon the bear was galloping quite fast, a sign that she was trying her best to close in on whatever she was chasing. She stopped suddenly about one-third mile from her cubs and fed on a calf carcass for almost 10 minutes. Then she started galloping back toward her cubs dragging the carcass, the legs and head dangling and getting in the way of her legs as she galloped, making several stops necessary to get a new hold. When the cubs saw her appear on top of the ridge, they galloped away far up a long snow slope. The cubs obviously did not recognize their mother. Down near the spot where the cubs had been digging, she dropped the calf and walked a half dozen steps toward the cubs that were some 75 yards away. She stood watching and possibly grunting. After watching alertly from the snow slope, the cubs advanced cautiously and tentatively, stopping to look, taking no chances. They remained cautious until they were quite near the mother, when they all fed together.

When the mother found the calf carcass after her chase, she apparently was motivated much as a human mother would be under similar circumstances. She was hungry yet worried about her cubs and wished to return to them. Her behavior was a compromise. Hunger took priority, and she fed, but as her hunger waned, her maternal instinct predominated and she hurried back to her cubs, taking the food with her.

On 1 June 1965 I watched grizzlies seeking caribou calves on the broad flats below Polychrome Pass. One bear had killed a calf and was feeding on it when another bear chased a nearby group of caribou. The feeding bear chased after the other. After a brief altercation in which the feeding bear was chased off, the second bear moved to the calf carcass, picked it up, and galloped off with it. After a few jumps, she dropped the carcass and galloped west about one-half mile where she was joined by a third bear, a 2-year-old cub. The female then led her cub at a steady walk back to the carcass and both fed. It was unusual that the female, upon finding the carcass, immediately fetched her cub before enjoying a meal.

Mistaken Identity

On 20 June, near the saddle of Sable Pass, two yearlings had moved about a 100 yards up slope from their mother. One was standing on its hind legs looking toward the opposite slope, a half-mile or more away, at another mother and her yearling. The two cubs were obviously apprehensive. When the second mother walked along on a contour of the slope, they followed, but on a higher contour as though wishing to maintain maximum distance from her. When their own mother moved up the slope, they kept well in advance of her, and watched alertly the opposite slope where the second bear and cub had been. This mother fed on year-old cranberries and farther up the slope she spent much of her time feeding on fieldmice and lemmings. Her nose informed her which hummocks were occupied, and these she ripped apart easily.

The two apprehensive cubs were soon far in advance of their mother. One of the cubs, light-colored and much smaller than its partner, seemed especially fearful and galloped ahead until it was some distance beyond the large cub that had stopped to feed. I do not know what happened, but the little cub seemed now to be afraid of its own mother. Having moved far ahead and probably having been out of sight much of the time, it perhaps became unsure concerning her identity, mistaking her for the other mother it had recently seen on the far slope. It made short gallops up the slope, to one side or the other, assuming alert, "scared" poses whenever it stopped. Its emotions seemed to keep building up, judging from its extreme exertions. After many dashes, the little cub moved far to one side of the mother, galloped to the bottom of the slope, and crossed a snowdrift a short distance from me. At this time it was almost a half-mile from its mother. Then the cub galloped up the slope to the other side of its mother, circled around her, and continued on nearly to the top of the slope where it repeated the short, fearful dashes, and seemed to undergo an emotional buildup of fear at each stop that caused it to gallop away suddenly. The behavior of the cub was not a game; its anxiety was too prolonged and too obvious for that. All this time the mother fed, unaware of the maneuvering of the little cub. She may have been aware of the large cub feeding some distance up the slope and assumed all was well. Eventually, the small cub maneuvered hesitantly down to the other cub, and at about the same time the mother galloped up the slope. Possibly the little cub had cried. The large cub moved toward the mother and the little one followed hesitantly. When it reached its mother, it sniffed her nose as though to make sure of her identity. The little cub's ramblings had taken about 1½ hours. The mother rolled over on her back and the two yearlings nursed.

On other occasions I have seen cubs behave similarly after being separated from the mother. On 20 September 1961 a yearling was feeding in a patch of berries some distance behind its mother and a second yearling. When he came within sight of her and his twin, he stopped to look but seemed uncertain of their identity. He made two or three short gallops to one side, looking questioningly at each stop, and then approached cautiously, stopping often.

Another yearling, busy digging for a ground squirrel, was left behind when its mother and twin moved on. After capturing and eating the squirrel, he hurried along the trail the rest of the family had followed and suddenly came upon them feeding in a green hollow 15–20 yards away. He stood at attention and uttered a snorting, questioning woof, three or four times. The mother did not look up; the grazing cub raised its head for only a quick look and resumed feeding. The uncertain cub relaxed and started to graze.

Feeding Courtesy

Bears ordinarily do not share animal food with others, even with family members, if supplies are limited, and other bears behave courteously and seldom interfere with a bear feeding on a morsel of ground squirrel or carrion.

On 2 August 1961, the smaller of two 3-year-old bears dashed down a gravel slope in pursuit of a ground squirrel that escaped into a burrow. The bear began digging vigorously and excitedly, jumping about the excavation spraying rocks between his hind legs. His companion watched from 4 or 5 feet up the slope. After about 15 minutes, the bear captured the squirrel and ate it while his companion watched quietly and made no move to interfere.

On 13 July 1962, a blond 4-year-old lay chewing something, perhaps an old bone. The brown twin walked slowly and apparently cautiously toward the blond and lay down facing it, its nose about 2 feet away. Soon, the brown one rolled over on its side and relaxed, but returned to its stomach and reached out with nose toward the blond, as though sniffing at what was being gnawed. The blond looked briefly at its companion who then rolled over in a puddle of water.

On 17 July 1959 the mother of two yearlings dug out a ground squirrel and captured it 6 or 7 yards from its hole. While she fed on it, taking small bites, one of the cubs grabbed the remains. This rarely happens. The mother struck at the cub with both front paws, a bluffing gesture and apparently an outlet for irritation, uttered a low growl, but permitted the cub to keep the squirrel. The cub moved about 7 or 8 yards and spent considerable time eating the remains of the squirrel. The other cub looked on without trying to obtain a share. The mother again dug briefly in the hole, then moved off and grazed. Previously I had not seen mothers voluntarily share their ground squirrels with cubs.

Retirement to Cliffs

During most of the summer, mothers with cubs, as well as other bears, rest day or night, wherever they happen to be feeding. But during the breeding period in May and June, family activity is somewhat different. At this season mothers with cubs climb steep slopes frequently and rest for the night on strategic ledges. Retreats are chosen away from beaten paths, as though for safety. Their only enemy would be other bears,

Fig. 24. During the breeding season females with cubs often spend the night in cliffs such as these near the Toklat River.

especially the males that at this season travel widely in search of a female. Males sometimes attack cubs, so perhaps cliffs are sought by females to protect their cubs. Below I describe some of my observations of mothers retiring to cliffs (Fig. 24).

Mother and Two Yearlings Seek Cliffs in Evening

On 18 May 1956 a mother and her two yearlings spent the day digging roots on the lower north slopes of Cathedral Mountain. In the evening they moved upward gradually until 9 p.m., when they stopped feeding and climbed higher up the slope to some cliffs and ledges. They lay down on a ledge and apparently spent the night there, for in the morning they were digging roots a short distance below their beds.

Mother and Yearling Observed in Cliffs on Igloo Mountain

On 5 June 1961, a mother with one yearling cub spent the afternoon digging roots on the Igloo Creek bar. At 7:30 p.m., I watched from my cabin as the bears climbed Igloo Mountain. The yearling led the way up a steep talus slope. In its exuberance it climbed every outcrop it encountered, and sometimes made little side trips for this purpose. Halfway up the slope was a slanted ledge and here the bears stopped on a grassy spot. The mother rolled over on her back and the cub nursed for 4½ minutes. She maintained her position rigidly for about 3 minutes after the nursing, then rolled over and appeared to be viewing the wide expanse of scenery below her—Igloo Creek, the north slopes of Cathedral Mountain, and the tundra reaching to the Teklanika River were all in view. At 8:30 p.m., the bears stood, moved about 20 feet, and lay down again. At 9:30 p.m., when I checked on them the yearling was nursing. I left them for the night and at 5:30 a.m., they were asleep on the same grassy ledge. When I checked at 6:35 a.m., the cub was nursing, and at 7:05 a.m., the female stood up, gazed over the country, and started down the slope.

In the evening I discovered these bears in a draw high on the slope. The mother was grazing in small patches of green grass on the south slope, and as she fed she moved slowly up a draw. About 9 p.m., they started to climb and at 9:10 p.m., reached a small grassy ledge on a sharp ridge, about one-half mile from where they had spent the previous night. They retired for the night at this spot and the following morning left the high ledge at 6:55 a.m., moving down to the swale where they had fed the night before. Thus, on two successive nights these bears had spent the night on a high, grassy bench away from any likely disturbance.

On 7 August 1961, the same mother and yearling were seen about 8:30 p.m., feeding on berries on a slope of Igloo Mountain. I watched them as they fed slowly up the slope until 10 p.m., when it was too dark to see them. I had hoped to observe them retiring for the night but they continued feeding on blueberries in the dark.

Mother and Yearling Retire on Ledge

Late in the evening of 22 May 1961 I saw a mother and yearling digging roots on a slope above Tattler Creek. The cub stood close beside or behind the mother, obviously wanting to nurse. In 10 minutes the female led the way about 300 yards up the slope. She drank at a creek for about 20 seconds, and the cub, for about 4. They crossed a small snowfield and stopped on a narrow ledge near the summit. Here, the mother rolled over on her back and the cub nursed. After the nursing, the cub lay close to her. Forty-five minutes later the bears were still on the ledge. The next morning when I went by they were digging roots on the slope a short distance below their beds. It seems certain that they had spent the night on the ledge.

Female and Two Yearlings High on Cathedral Mountain

On 30 May 1962 about 7:30 p.m., a female and two yearlings (one crippled) were sleeping on the talus at the base of a cliff near the top of Cathedral Mountain. They were still there when I looked at 11 p.m. At 3 a.m. they were resting 50 yards from the bed, and at 5:30 a.m. they were digging roots 25 yards from where they had been seen at 3 a.m. Apparently, they had risen a few minutes earlier.

Mother and Two-Year-Old Climb High in Cliffs

On 31 May 1962 a mother followed by a 2-year-old cub, fed on roots near the base of the north end of Cathedral Mountain during the day, stopped feeding at 9:30 p.m., and moved on a contour to a canyon. In one place the mother stopped for 10 minutes to dig roots before continuing on her way. I lost sight of them when they disappeared into a canyon around a shoulder, but a little later saw them climbing a long, steep talus slope among sharp pinnacles. They climbed steadily. Some ewes with lambs that had retired for the night in this rough country moved a short distance out of their way but were scarcely noticed by the bears. When the bears neared the top of the mountain, they were hidden among the pinnacles. Apparently, they had climbed high to bed down for the night. In the morning they were feeding again where they had fed the previous evening.

Mother and Two-Year-Old Seek Cliffs

Late in the afternoon of 23 May 1959, a mother and her 2-year-old cub stopped digging for roots and started to climb. They rounded a shoulder of the mountain and continued up the steep talus slopes of a canyon. At times they were hidden by the numerous outcrops and were last seen near the top of the mountain. They apparently were going up high for the night. In the morning they were back near the base of the mountain, digging roots where they had fed the previous day.

Mother and Two Two-Year-Old Cubs Spend Two Nights in Same Cliffs

On 24 May 1963 I saw a mother with two 2-year-old cubs climb to a high ledge on Cathedral Mountain at 4 p.m., after they had spent the afternoon feeding on roots on lower slopes. They still were resting at 5:20 p.m. By 6:50 p.m., remaining high, they had traveled one-half mile around a shoulder, climbed a steep slope, passed over some cliffs, and dug a platform on a steep slope. The cubs nursed at 7:50 p.m. and then they all lay down on the shelf. The next morning at 4 a.m. the family was still resting in the same spot. Later in the day, they were digging roots where I had seen them feeding the previous day.

On the evening of 25 May at 6:25 p.m., I saw the same family resting close together on a bench high on Cathedral Mountain above where they had fed during the afternoon. The following morning at 7 a.m., they were seen coming down from the cliffs.

On 30 June, about 9:30 p.m., I saw this same family on the north side of Cathedral, climbing high on the slope, They were nursing at 9:55 p.m., after which the mother lay down on the slope, one cub near her, the other feeding 75 yards away. At 10:15 p.m. one cub rested 10 feet from the mother, the other 75 yards away. The following morning at 5 a.m. the family was resting together on a prominent outcrop, a little above where I had left them in the evening.

Mother and Single Two-Year-Old Cub Spend Two Nights in Cliffs

On 25 May 1963 a dark mother and one blond 2-year-old cub climbed a long slope and at 6:30 p.m. reached an outcrop above Tattler Creek. After nursing, they rested. The following morning at 6:30 a.m. they were still in their beds and remained there until 8:45 a.m. They then traveled across a slope and around the far side of a ridge. When crossing a snowfield, they started a considerable snowslide. They galloped across another snowfield, sinking and sliding, and disappeared over a small side ridge. That evening the bears were resting 200 yards from the ledge used the previous night. I was unable to visit them the following morning, but presumably they spent the night on the high slope.

Mother and Two Two-Year-Old Cubs Retire Early

On 30 May 1964 a mother grizzly and her two cubs were digging roots industriously on a slope of Cathedral Mountain. About 4 p.m. the grizzly mother climbed up among some rugged outcrops. When I checked on the family at 8:15 p.m., the mother was still resting on the outcrop and the cubs were digging roots nearby. At 4:30 the next morning, the family was resting, and for 15 minutes there was no movement. Then the mother raised her head a few times for a brief look, and at 4:50 a.m. she stood up, walked a few steps, and gazed at the landscape below her for a minute or two before moving down the slope to resume root digging.

Mother and Two-Year-Old Cub Spend Night in Low Country

On 18 June 1965 at 7:30 p.m., I discovered a mother and her 2-year-old cub resting on a flat near an open stand of tall willow brush. During the next hour, each bear walked a few steps to leave a dropping and returned to its bed. Part of the time the cub rested with its back against the mother. They lay in various positions, on their sides, on their stomachs, with hind legs stretched out behind, and on their backs. The following morning 11 fresh droppings were found near the beds.

On 11 July 1965 I saw the same family climb a low bank above the west branch of East Fork River about 9:10 p.m. The cub nursed and then the bears apparently settled down for the night, because at 4:30

a.m. I saw them walking away from the beds to graze on the river bar. On these two occasions they had not retreated to cliffs for the night, although the night of 18 June was still in the rutting period.

Cliffs Sought in Daytime

Besides retiring to cliffs at night, mothers, especially those with spring cubs, and young bears seek high country and cliffs for escape from danger during the day. On 19 July 1953 two photographers surprised a mother with a spring cub. She was about 50 yards away and faced them with head down in a stiff posture, but she soon relaxed and lay down for about 10 minutes with the cub between her paws so that only its head showed. Then she led the cub toward a pond just south of Cathedral Mountain. When she caught sight of a large bull caribou ahead of her, she ran back toward the cub as though to shield it, then led the way, galloping to the mountain slope and continued to the top. She probably had not identified the caribou—perhaps she thought another human was approaching.

On 18 June 1953 I watched a young bear, perhaps a 3-year-old, grazing on grass, horsetail, sourdock, and the herb *Boykinia*. When he neared a pole he used it as a back scratcher, twice standing on hind legs to rub. He sniffed at a few ground-squirrel holes, then climbed some distance up the mountain and lay down on a prominent lookout point, apparently for security. An old bear usually will lie down to rest wherever he happens to be, but this young bear sought a point from which he could watch his surroundings.

On 20 June 1953 three of us came upon a mother bear with three spring cubs near the south end of Cathedral Mountain. Upon seeing us, she took her young family over the top of the mountain. Later in the day, and farther north along Igloo Creek, I discovered a mother and a spring cub. She led the way up a long cliffy slope on which climbing was difficult because the terrain was steep and the gravel loose. The cub climbed more easily.

On 2 June 1955 a mother and two spring cubs were below this same slope. She climbed the steep, gravelly incline, as had the mother 2 years before. One cub started sliding, but turned so he faced uphill and managed to stop by digging in with all four feet. After climbing almost to the top of the high steep slope, the mother recovered her composure, moved along a contour, and gradually worked her way down again to the creek bottom.

On 4 June 1955 a mother with two yearlings, after feeding on a caribou carcass, climbed up among cliffs to a point one-quarter mile away and rested where she probably felt secure. However, the following day, after feeding on the carcass, she rested near it.

On 30 May 1959 about 5:30 p.m. a mother with a single 2-year-old cub hurriedly left a mother and two 2-year-olds and a set of large, twin bears that were 300 to 400 yards away on the Toklat River. The mother and her cub climbed two-thirds of the way up the northeast slope of Divide Mountain, stopping only when they were among the cliffs. A few minutes after they stopped on a ledge, the cub nursed. I watched them resting for 1 hour and 40 minutes after the nursing, and left them at 7:30 p.m. It seems likely that this family spent the night in the cliffs.

In 1962 I observed on several occasions 3- or 4-year-old cubs resting alone during the day in the rocks above East Fork River.

In the early morning of 3 July 1965, I saw a lone bear, a mother with two spring cubs, and a mother with one yearling at the base of the north slope of East Branch Range. The lone bear and the mother with spring cubs were perhaps 150 yards apart, the other family 300 yards away over a rise.

Later, I saw the lone bear and the mother with spring cubs climbing the steep slope about 250 yards apart, a deep draw between them. I had not seen the start of the climb and do not know what instigated it. Perhaps the bears startled each other and each sought safety in the cliffs. As the bears climbed, they watched each other but the lone bear climbed more rapidly. Far up the steep slope the family crossed the draw below the lone bear. One of three snowfields they crossed was so steep that one of the cubs inadvertently slid about 40 feet before he was able to face upward and stop the slide with his claws. The family disappeared among the outcrops, and the lone bear climbed to the top of the ridge where later I saw him resting.

Thus, bears, especially families in spring and early summer, seek resting areas that offer a good view of the surroundings.

Mother–Cub Separation

The mother–cub association lasts over 2 years, much longer than has been supposed. The cubs remain with the mother for 2 full years and for at least a part of the third. Occasionally, a single cub may remain with its mother for 3 full years and a few months into the 4th year.

Of the 69 mothers followed by 2-year-old cubs that I have recorded, 30 still were followed by their cubs after 1 July. Of these 30 families, 11 were known to be intact in August, and 8, in September. Five mothers were followed by cubs over 3 years old, and two other 3-year-old cubs were near the mother under special circumstances. Several of the families that I failed to see after 30 June probably were intact in July and later.

It may be significant that, with one exception, only single-cub families were intact when the cub was over 3 years old. It is logical to assume that the single cub is most likely to remain with its mother in its 4th year because cubs seek companionship and a single cub has only its mother.

On the other hand, twins and triplets early become rather independent in their play and companionship, so that for them a break from the mother may be easier.

Various incidents pertaining to the family breakup will be described to show some behavioral characteristics at that time.

An Early Separation of Cubs Due to Breeding

On 18 May 1960, as I was starting to climb a ridge on Cathedral Mountain about 1 mile north of Tattler Creek, I caught a glimpse of three bears climbing to cross the base of the ridge I was on. They were about 100 yards away and coming directly toward me. I retreated to my car which was parked on the road near the adjacent creek. In a few minutes the bears appeared, moving forward methodically in single file, the mother leading, and the two sturdy 2-year-old cubs following closely. They crossed the base of the ridge where I had been, crossed the creek, and came directly toward me. When they were 50 yards away they saw me, looked, turned slowly, retreated a few yards, climbed a steep bank, and dropped down on the other side. I recognized the family as the brown mother and one blond and one dark cub that I had known the previous year. Now the cubs were about 2¼ years old. They crossed Igloo Creek and climbed a slope. Here, they fed on the previous year's crowberry crop for about 10 minutes, then continued one-half mile to another slope much favored by bears at this time of year. All began turning over sod on the slope to feed on the roots of the herb *Hedysarum*. At 11:45 a.m., 1½ hours after they were sighted, the mother lay down and the blond cub moved to her side, but moved away in a few moments to continue digging roots. In 3 minutes the mother stood up, took a few steps, and lay on her back in nursing position. Both cubs hurried to her and nursed for 4 minutes. Then she turned over, moved a short distance, lay on her stomach, and the cubs rested against her, one on each side. At 12:40 p.m. the mother stood up briefly then lay down again, rolled over on her back, and the cubs nursed for 5 minutes. Five minutes after nursing, the cubs went to feed on roots. At 1:30 p.m. one of the cubs returned to the mother, who was still resting on her side, and pushed his head under her arm, trying to nurse. In a few minutes the other cub came over and the mother obligingly turned over on her back and a 4-minute nursing ensued. Three nursings in less than 2 hours! At 1:50 p.m. the mother moved southward, crossing slopes and draws and Tattler Creek as she proceeded to Sable Pass, without loitering along the way. Moving westward on Sable Pass to a point near the base of Sable Mountain, the bears stopped at intervals to feed on berries. About 5 p.m., as the cubs continued to feed, the mother moved forward and was soon about a quarter mile ahead. Just before dropping into a deep ravine she looked back for the cubs who were galloping toward her. As soon as the cubs reached her, she reversed her direction and started eastward at a gallop. Apparently something attracted her attention when she stopped to look back toward the cubs. She hurried, alternately galloping and walking rapidly, stopping a couple of times to look back at the cubs who were following a little distance behind and I heard her cry sharply, as though urging them to hurry. Next, I saw an eagle circling low and almost alighting on a steep, bare slope ahead of the bear. A little later a cow caribou ran down this slope and, as she passed, the cubs made a short dash toward her. The mother bear galloped up the steep slope to the spot where the eagle had hovered and picked up a dead, newborn calf caribou. The head and legs of the calf dangled from her jaws as she galloped down the slope to more gentle terrain.

The feeding behavior was interesting. The mother lay on her stomach facing downhill as she fed on the carcass and a cub was tugging on either side, at right angles to the mother. After 15 minutes the carcass was dismembered, and each cub moved a short distance away with a sizeable piece. The mother ate what remained, sniffed about a little, then approached the blond cub slowly and warily, and watched him as she lay crouched on her stomach for almost a minute, a few feet away; then she made a sudden pounce on the cub's piece of carrion. The cub drew back with some sharp cries but managed to retain some of the meat. When the mother had devoured her stolen morsel, she went up to sniff at the spot where most of the calf had been eaten, then repeated her pilfering maneuver, grabbing the remains from the cub. At first, the calf carcass was common property, but after the bears had separated, each developed a sense of property that was recognized by them all.

At 6:20 p.m. the mother rolled over on her back and the blond cub nursed. Three times the cub stopped nursing to lick the mother's face. I guessed that it was licking blood from the fur. In 4 minutes the mother rolled over on her side and when the cub persisted in trying to nurse, she rolled over on her stomach. About this time the dark cub, 20 yards up the slope, finished his piece of calf carcass. He had missed the nursing. When I left at 6:45 p.m., the mother was lying on her stomach and close on either side was a cub also on its stomach.

I relate these observations to show the intimate relationship that exists in the family so close to the time of family breakup.

Two days later, on 20 May, I met the two cubs on the west side of East Fork River, about 3 miles from where I had left them with their mother on the 18th. The two cubs dropped down to the broad, gravel river bar, crossed several channels of the river, and fed on roots along the east side. After 2 hours of feeding, they started to re-cross the bar. Midway across, they were attracted by something up the river and one stood on hind legs to watch. Then both started galloping westward. Soon a huge, dark, extremely fat male grizzly came into view. He was so large and fat that he could manage only a laborious shuffling trot. The cubs climbed the steep slope, keeping about 100 yards or less ahead of the male, stopping frequently to watch his progress. The male gave up the pursuit when he reached the ridge top. The two cubs were already managing on their own very well.

On the following day, 21 May, the cubs were in the same locality. Just below the East Fork bridge they dropped off the ridge, galloped across the river, and continued down the bar to the entrance of a narrow draw, just beyond a bluff. Here they received a scare, for they galloped back the way they had come. They climbed 50 or 60 yards up the steep face of the ridge and lay down for 15 or 20 minutes, watching with strained attention the mouth of the draw where they had been frightened. They moved a few yards further on out of my view. Their interest in the draw from which they had fled suggested that their mother was there and very likely with a male. I watched the draw for 2 hours but did not see a bear emerge. I assumed that if the mother were consorting with a male I would see her often in the area in the following days, so I departed without investigating. The next day two photographer friends who had known the family saw a pair of bears at the East Fork bridge, and their description indicated that the female was the mother of the cubs. I saw the mother alone on 27 May. If she had mated the honeymoon was short, but long enough, for it is likely that she was ready to breed at once when she left the cubs. (The following year she had two cubs in the spring.) The mother and the 2-year-old cubs remained in the Sable Pass area all summer, but mother and cubs were never seen together. On three occasions she was seen about one-half mile from the two cubs, but too far for them to be cognizant of each other. The early separation of this family appears to have been due to the mother coming in heat.

On 11 June 1960 two other 2-year-old cubs were seen alone along Igloo Creek. They also probably had been deserted by a mother in heat.

Drifting Apart

In 1940 three robust 2-year-old cubs followed their mother throughout the summer. It is doubtful if this female mated for during the breeding season the family was seen at short intervals. On 17 and 18 September the family was still intact, but on 23 September the three cubs were fully one-half mile from the mother. They may have rejoined her later, but the relationship had been very loose for some time, and it is likely that they had drifted apart.

Late August Separation

On 21 August 1956 a mother and two dark 2-year-old cubs were feeding on buffaloberry a mile beyond Toklat bridge. The following day the mother was feeding about one-half mile from the two cubs. On 24 August the female was not seen and the two cubs were moving about together, apparently on their own.

Early September Separation

In 1959 I observed a family in the same buffaloberry area west of the Toklat River where I had observed a breach in 1956. The circumstances were much the same. On 4 September the cubs were feeding with their mother. The following day they were far up a draw near some sheep that were keeping an eye on them. Later, the cubs came down and crossed the road near where three of us were standing. A bear that appeared to be the mother had been feeding near the stream when we first arrived and she later moved over a rise a mile away. On 7 September the two cubs were together with no other bear in sight. Apparently, the cubs were on their own, whether due to the mother's antagonism or to a mutual loss of attachment between mother and cubs was not learned.

Separation as Result of Mutual Inclination

In the summer of 1961 I observed a blondish female and two rather dark 2-year-old cubs on 26 different days beginning on 25 June. On 25 July, when I spent several hours with this family, I saw nursings at 9:30 a.m. and at 2:50 p.m. It seemed to be a rather cozy family. On 26 August the mother and cubs fed together on friendly terms, the cubs feeding close to the mother and also 100 or 200 yards from her. On 27 August I watched the cubs feeding together for 4 hours. Later, I discovered the mother almost a mile away. When last seen, the cubs were feeding eastward and the mother was moving west. On 28 August the cubs again were feeding together about one-half mile from the mother. On 1 September only the two cubs were seen moving about together. Apparently, a separation of the mother and cubs had taken place as a result of mutual inclinations. No antagonism was noted.

Intolerant Mother Causes Family Breakup

In 1960 I watched frequently a well-marked mother and two 2-year-old cubs that I had observed many times the previous year. The family made its first appearance on 30 June 1960 and was seen every day but one until 12 July. During this period a larger bear that appeared to be a male stayed near the family on rather familiar terms, as though a breeding period were in the process of terminating (*see* section on mating). After 12 July, I did not see these bears for a month; apparently, they had moved a few miles southward.

On 11 August, when I discovered the family about 5 miles west of the top of Sable where they had been seen last on 12 July, the group was breaking up; even the two cubs went off in different directions. I first noticed the blond cub as it fed in a semicircle. The female was feeding about 300 yards from this cub and 50 yards from the very dark cub that was standing, watching its mother. I guessed that it was in disfavor. After a time, it fed within 30 yards of her and disappeared behind a slight rise. In about 15 minutes the mother moved behind the same rise, and the dark cub emerged from the opposite side. He moved about 70 yards away and lay down, apparently retreating from the mother. A little later the blond cub moved toward the mother in its feeding and disappeared behind the same rise, but at once emerged with a rush. Apparently, it had been threatened by the mother. This cub alternately galloped and walked eastward about 1½ miles. At the same time the dark cub started westward at a fast walk and was last seen about 2 miles away. The mother remained feeding. On the following day the mother was feeding one-half mile to the east and the blond cub was near the spot where first seen on the previous day. This cub was seen here again on 16 August; on the 17th the mother was in the area and the blond cub was about 200 yards away. The blond cub approached quite near a draw the mother had entered, but later galloped away as though threatened. What appeared to be the dark cub was traveling along foothills to the south, a mile or more away. On 18 August the mother and the blond cub were seen in the area about one mile apart. The mother had been the aggressor in causing the separation.

Late September Separation of Mother and Two-Year-Old Cubs

Sam Woolcock told me that late one summer he watched a mother followed by two large cubs that he thought were 2-year-olds. Suddenly the mother growled and threatened the cubs who galloped away into the distance. It appeared that this mother no longer tolerated them. The surprising element which Woolcock pointed out was the long retreat of the cubs and they apparently had no intention of returning. The behavior of these cubs was similar to the behavior of those in the incident I described above.

Breeding Mother Antagonistic to Old Cub

On 9 June 1955 a large male and a female were consorting between the forks of the Toklat River. Both bears were digging roots. Near them was a small bear that I judged to be a 3-year-old. It dug roots too, but most of the time it just stood and looked toward the other two bears. After a time, it circled downwind and walked slowly to within 25 yards of the female. She made a short charge of about 10 yards, causing the small bear to gallop away. Later, the cub moved away slowly from the male who was feeding toward it. The next time the cub approached the female she made a determined charge of about 100 yards. The cub persisted and a third time moved close to her and stopped 35 yards away, watching as though wanting to join her. When the male in its feeding moved gradually toward the young bear, it retreated slowly a short distance. Apparently the mating session was breaking up the mother–cub relationship.

Breeding Female Tolerating Old Cubs

On 27 June 1961 I watched a pair of breeding bears whose behavior differed from that in the incident just described. Two large 2- or 3-year-old cubs fed in the area, sometimes within 100 yards of a pair. The male was observed mating with the female for a prolonged period. Later, one of the cubs approached the female until they stood facing each other, with noses only 2 or 3 feet apart. Apparently the cub was still on friendly terms with its mother. As the male moved slowly toward the mother and cub, the cub walked away for about 200 yards. Two days later the male and female were one-quarter mile apart, with one cub feeding about 100 yards from the female. This was late in the breeding season which suggests that the mating of the pair had terminated. These bears were not seen again so it was not learned whether the cubs resumed a close association with the mother.

Cubs Still With Mother When Over Three Years Old

In 1961 a very blond mother and yearling were seen at intervals on Igloo Mountain and seen frequently there in 1962 when the cub was over 2 years old. On 11 August 1962 I watched the two bears cross several ridges and draws as they traveled toward the Big Creek side of the mountain. The cub led the way. He was the restless one and the mother followed compliantly. On one occasion the cub was two ridges ahead of the mother, who occasionally tarried to feed in a draw. Once the cub waited briefly until she came into view over a ridge, then continued on his way. The same behavior was noted on 14 August. The mother was not indifferent to the cub, for she followed him even though his restless behavior was unusual and apparently different from her own tempo. The two bears were last seen on 11 September. When the cub was 3 years old, I saw mother and cub in the area on 30 and 31 May 1963, still on friendly terms. Their unusual blondness made misidentification unlikely. The two were not seen together after 31 May, but a week later I saw what appeared to be the cub. Probably the female moved away to breed. Apparently they had hibernated together.

Mother and Two-Year-Old Close Companions in September

It seems likely that a 2-year-old may hibernate occasionally with its mother. In 1962 a mother and 2-year-old cub that I had been observing for 2 years were still associated closely when last seen on 12 September as they left Sable Pass for the Teklanika drainage.

An Old Cub Chased by a Mother With Spring Cub

On 14 July 1953 I watched a mother with a single spring cub chasing away a large cub, probably a 3-year-old, that persisted in remaining near (Murie 1961). Once, when the mother rushed at the big cub, she overtook him and bit him severely on a hind leg. The big cub had followed the mother and spring cub all spring judging from the familiarity that existed among them. The spring cub had no fear of the larger cub. But on this day the mother's tolerance had apparently reached a limit. It appeared that the mother had mated while being followed by a 2-year-old cub the previous year and that the cub had remained the rest of the summer and hibernated with her. The coming of the new cub caused unusual complications (Figs. 25, 26, 27).

Mother and Three-Year-Old Cub Together

On 5 September 1965 I watched a mother and her single 2-year-old cub near the Toklat River foraging for buffaloberry and digging briefly for roots. The berry crop was a failure so the two bears wandered widely as they searched for berries among the willows. They became separated frequently but always sought each other when this happened. The companionship seemed as close and intimate as ever. I knew this family well, having watched them many times during the three summers the mother had been abroad with this cub.

Fig. 25. Mother with spring cub, followed by a 3-year-old cub.

Fig. 26. The old cub watches mother, not daring to approach. Earlier she had charged, bit him severely on a hind leg, and stood over him.

Fig. 27. The older cub, chased by the female, had approached close to her and is here leaving in a hurry as she growls threateningly.

The following day I returned to look for the family and found them on Highway Pass; they had moved over 2 miles farther west since the previous evening, and were travelling at a fast walk. A few times the female broke into a lope, and I could see from her alertness and general behavior that she was hoping to surprise a ground squirrel. She stopped at a set of holes, dug for a few minutes, then loped forward and came upon another set of holes which, after a little digging, yielded a squirrel. The cub who had tarried at the first set of holes to continue digging, captured a squirrel about the same time. The mother moved over a rise, but reappeared and returned 150 yards to her cub who was finishing his squirrel. The female then led the way as they loped toward Slide Lake. They stopped for a few moments at some bushes, apparently a few buffaloberries, then moved forward steadily along Slide Lake and disappeared north of it. The companionship and solicitude exhibited by the mother at this late date suggested that she would hibernate with her cub, and that the cub still would be with her when it was over 3 years old.

In 1966, the following spring, the family was seen on the Toklat River where it often was seen during the previous 3 years. It was seen on 30 May, 3 and 4 June, and was reported on 12 and 13 June. The mother and 3-year-old were on friendly, intimate terms when last observed.

Summary

No doubt a variety of factors cause the variation in timing of family breakup. The onset of the breeding season when cubs are 2 years old is associated frequently but not always with separation. Some females either do not come into breeding condition, are not found by a male or perhaps resume association with their cubs after breeding. In most cases the mother plays an active, aggressive role in terminating her association with her cubs. Sometimes, especially in litters of two or three, the cubs drift away from their mothers of their own accord.

Cub Companionship After Separation From Mother

After twins or triplets separate from their mother they generally continue to associate. Over a period of years, I have seen over 50 sets of twins and 3 or 4 sets of triplets continue their companionship. One set of twins remained together for three summers after their mother left them as 2-year-olds.

Brown Female's Cubs

In 1959 a female was followed by two well-marked yearlings, one dark brown and the other blond. Both cubs appeared to be females. The family was observed, confined closely to Sable Pass, from mid-June until early August. In 1960 the two cubs were on their own on 20 May. When these cubs were yearlings, the smaller, blond cub was more active, always straying more widely when feeding, and this restlessness continued after the two cubs were on their own. These cubs remained together on the East Fork River bar from 20 May to 30 May. On 31 May the blond cub moved a mile up the river bar. On 5 June they were still a mile apart. By 19 June the dark cub was near the top of Sable Pass, 3 miles to the east, but the blond was not seen. On the 26th both cubs were seen near the top of Sable Pass about one-half mile apart, but on the next 2 days only the dark cub was seen. On 7 July the two were together near the top of Sable Pass, and for the remainder of the summer, until 26 September, the last day I observed them, they always were seen together. During this time they were seen at short intervals on 27 days.

When together, the bears were always chummy, although I did not see them play together more than two or three times. They frequently rested, touching or only a few inches apart. On 7 July, after they had fed steadily on dock for one-half hour, they lay down side by side. The one lying slightly farther back moved a few inches forward, then a few more inches until its nose was even with that of its companion. The blond cub remained much more active all summer, seemed always restless, and was generally in the lead when they fed or traveled.

On 9 May 1961 the two cubs, now 3 years old, were seen along Igloo Creek 2 miles from Sable Pass. The following day they were a mile apart. I did not see either bear again until 23 May, but from then until 15 June I saw the dark cub nine times in the East Fork River area. Between 11 and 17 July the cubs were together, ranging from the top of Sable Pass to a point about 6 miles to the west on Polychrome Pass, where they were seen on 17 July. While the blond fed in a green swale, the dark cub moved over a rise. Half an hour later, when the blond was leaving the swale, the dark cub returned to meet it. They touched noses, rose up on hind legs to hug and wrestle briefly, and walked away over the top, the blond in the lead. Between 18 July and 18 September the twins were seen together 22 times and apart (up to 2 miles) 9 times. Frequently, one bear was left far behind temporarily in their feeding activities. On 17 August I saw the dark cub follow a trail of the blond for over one-half mile. The blond saw its partner approaching from a distance and recognized it, for it resumed feeding at once. From the middle of August until last seen on 18 September the cubs were usually together. On the last day they were feeding down Igloo Creek toward the place where they were first seen in spring.

In 1962 the blond cub was first seen on 2 June along Igloo Creek. The brown cub was seen on the East Fork bar on 15 June and on Sable Pass on 24 June. The two big cubs were observed together on Sable Pass 18 times between 3 July and 3 August. On 23 August they walked up the bar of the East Fork River, the blond leading the way. They were not

Fig. 28. Twins, about 4 years old, still companionable after leaving their mother.

seen again in 1962 and were not recognized in 1963. In 1962 the cubs were 4½ years old, and both showed more maturity in their actions. The cub-like quickness of movement was gone and the gait more deliberate. After they came together on 3 July, they remained associated closely, often feeding only a few feet apart and resting close together (Fig. 28).

Other Twins Together

On 24 July 1963 I noted a large, dark cub with a crippled left front foot feeding on Sable Pass. When walking, he carried the injured foot up and his movements were very restricted. He was alone on the following 2 days, but on the 27th a blond cub, slightly smaller, rested 6 or 7 feet from him. After a time the blond stood up, nosed the cripple, and moved away as it grazed. The cripple hopped toward the blond, who, on seeing him, returned and played with him for 25 minutes. The cripple was handicapped in the play by his bad foot. The blond was the aggressor and for much of the time was on top of the cripple. Later, the cripple was on top, and with a firm hold on the blond's neck, shook him vigorously. The blond obviously enjoyed this pummeling as it lay on its back, relaxed. Then they stood on hind legs wrestling. This phase of the play was most difficult for the cripple, and apparently he did not enjoy it for after a time, the cripple stiffened in his attitude and the blond withdrew. Later in the day, they fed one-half mile apart. On 28 and 29

July they were together, the blond feeding nearby and then far off, very restless. The following day only the cripple was seen. They were together on 1 August; on 6 August the cripple was 2 miles down Igloo Creek, apparently searching for berries, his foot much better. On 9 August they were together again on Sable Pass, the last time I saw them in 1963. These twins behaved similarly to the brown female's twins; that is, they continued to associate but often moved apart and were out of contact for a day or longer.

Twins Split Up

On 11 August 1960 a mother separated from her 2-year-old cubs (*see* Mother–Cub Separation). When threatened the dark cub traveled at least 2 miles west, and the blond traveled east 1½ miles. On 12, 16, 17, and 18 August the blond cub and the mother were in the same general area. The dark cub was seen a mile to the south on the 17th, traveling southeast. Whether the two cubs rejoined each other was not determined, but it appeared that they were not seeking each other. The blond was considerably larger than the dark one, and I had noted earlier in the summer that the blond played so aggressively that the dark one often tried to escape. This background suggests that the dark cub may not have been anxious to continue the association.

Two Companions Call to Each Other

On 25 August 1949 I startled two bears in the woods along the Teklanika River. One ran into the woods above the road and the other ran below the road. One started to utter chuckling, baby-like sounds that were answered by the other bear. They called and answered three or four times before the bear on the lower side circled to join the other. This is the only time I have heard cubs call to each other. They appeared to be 2- or possibly 3-year-olds.

Companionable Behavior of Three Cubs

On 13 July 1962 on Sable Pass I watched the behavior of four 3- or 4-year-old cubs for several hours. Three of them maneuvered about together and may have been triplets; the fourth rested 200 yards from the others. He was the most inactive young bear I have ever seen and rested during the 8 hours I watched. A dark-brown cub was the aggressor in play with a blond cub. He kept pushing in toward her, finally taking briefly a breeding position. Soon after this, the play broke up, the dark bear walked to a tan cub and stood beside it, the blond also moved close, and all fed for a time and later rested. A few hours later the brown and tan were traveling together down the slope and the blond followed. When they reached a snowfield, the brown chased the blond, who retreated at an easy lope. When the brown walked away, she followed. Soon all three moved up the slope again, the two dark ones together and the blond just ahead of them. The dark one followed the blond at intervals. The two brown bears were together often but did not play. Perhaps the blond was a female and therefore attractive to the teenaged brown which had the appearance of a male.

The blond at one point lay on its belly chewing something, perhaps an old bone, The dark-brown, ever interested, moved to the blond, and lay down before it on its stomach with head resting on the ground so that its nose was only 2 feet from what it was chewing. He then rolled over on his side, back on his stomach, and pushing nose forward sniffed at what was being chewed. Bear etiquette apparently made him behave properly. (Cubs often watch stoically while their mothers feed on ground squirrels.)

The three then went higher on the slope to a long snowbank. The blond and the dark bear played. The blond, having the uphill position, dominated the wrestling match, until the brown galloped away, followed by the blond. A little later all three disappeared from view.

Why the tan cub did not play and why the fourth bear remained aloof were mysteries. The latter may have been hurt slightly in rough play. I saw these bears frequently, often scattered, but never knew their relationship.

On a few occasions I have seen a lone cub chase a smaller lone cub so earnestly that the small one traveled some distance after the pursuit stopped, as though wishing to remove himself from the area, at least for a time.

To what extent unrelated cubs mingle after leaving the mothers was not determined, but general observations indicate that such cubs remain apart. Solitary life is so typical of adult bears that one would expect young, lone bears to become accustomed to it.

Mock Fighting

On 3 July 1948 two bears that I judged to be 2 or 3 years old were facing each other about 10 feet apart when discovered. The darker one moved backward in slow motion, one leg at a time, up a bank. Once at the top, he moved away slowly, keeping one side toward his immobile companion. They behaved as though hostility existed between them. As the dark bear began grazing away, the light one galloped toward him, and the dark one snarled. Both bears growled with jaws open and close together. The light one closed in and seemed to bite the neck of his companion. They separated, stood watching each other, then both grazed for a time, moving about without trying to separate. When a hundred yards apart, the light bear again galloped to the dark one who turned and growled. The light one stood still with nose almost touching the ground, a sort of on guard pose, and the dark one soon lay down on its stomach until the light one walked away. While the light one started digging for a ground squirrel, the dark one walked rapidly away. Seeing this retreat, the light galloped after its companion who snarled and again lay down. The light one circled, and soon they both fed toward their original position. Later there was another chase, and after a time the dark bear walked off some distance and was not followed. The two bears had behaved in much the same manner as a mated pair. The following day they were seen about one-half mile apart.

On 14 July I saw the light bear gallop toward the dark one and soon overtake him. They faced each other a few feet apart with heads down, noses almost touching the ground. A little later they stood side by side, heads still down. The light bear edged slowly away and then the dark one chased after it and soon they faced each other again, the light one lying down part of the time. The dark one backed away slowly, then walked off. Later the bears, now some distance apart, were seen rolling on their backs, legs pawing the air. They had another chase. Three or four hundred yards up the slope was a lone, larger bear who later came down the slope in her feeding and chased the dark one, who stopped after a gallop and the two faced each other. The dark one lay down on its stomach and the big bear moved off. This third bear was large enough to be the mother but it may have been an older cub.

On 25 July these three bears were seen again in a green swale below a snowbank. The two smaller bears chased each other in play, then all three grazed in an area 50–75 yards across. The two smaller bears climbed onto the snowbank and for over one-half hour wrestled, mauled, and mouthed each other. When they moved off the snow, one climbed a 6-foot boulder and, from above, sparred with the one below. Then both were crowded on the rock. They wrestled but there apparently was a truce about shoving off the rock. This play continued for about 10 minutes after which they wandered over the skyline to the south. The large bear, possibly the mother, fed northward. This might have been a family in the process of breaking up.

Summary

After separating from their mother, twins and triplets frequently continue their companionship for a while. However, even when relationships remain friendly, the animals often wander alone. Such amicability and tolerance can extend at least to 4½ years of age and possibly longer.

Fig. 29. A lone bear seeking food after an early September snowfall. Soon it would den up for the winter.

5
Subsistence

Hibernation

Hibernation is correlated with the period of food scarcity. When autumn snows arrive, the bears continue for a time to dig roots, to excavate ground squirrels, and, in places, to feed on berries. But as autumn progresses, the ground freezes, berries become buried by snow, and general feeding conditions deteriorate. The daily regime of gorging over several months results in warmly furred bears prepared to wait out the winter months in underground chambers. The denning period, judging from general observations, extends from late October and early November to April (Fig. 29).

On 11 October 1939 a lone bear was observed digging a den on a high, rather steep slope. A foot of snow lay on the ground at the time. The following spring, on 29 April, I had my first view of the den. Three fresh trails led out from the den over the snow, indicating that the bear had been visiting or occupying it recently. The thin sod roof over the chamber caved in during the summer. The chamber was about 4 feet from the entrance and about 5 feet in diameter.

Years ago (29 March 1922) my brother Olaus, traveling by dogteam, stopped at the Knight Roadhouse down the Toklat River several miles north of the park. The story of a bear encounter related to him at the roadhouse included the information that a bear was digging a den in November. The story is of interest and I shall quote from Olaus' diary:

> Sam Federson told me about his encounter with a bear last fall and gave me the skull. On November 6, 1921 Henry Knight saw a bear in the distance above timber near Chitsia Mountain and shot three times. The bear was wounded slightly and ran into his den nearby. The bear had been out gathering for a bed in his den when shot at. Next day Sam Federson and Henry Knight returned for the bear. They found that the bear had left the den and run down below timberline. There was considerable snow on the ground and the men were traveling on snowshoes. It was ten degrees below zero. They were going slowly through underbrush. Sam was ahead with his mittens on, when suddenly he was confronted by the bear, and before he could use his gun the bear hit him in the

head and knocked him down. The bear came after him but he put up his feet against the bear and the bear shattered his snowshoes then came after him again, this time seizing his arm. Knight shot him in the heart and the bear fell on Sam.

Sometime later I obtained measurements of this bear's skull from Richard H. Manville. These measurements indicated that the bear was a large male.

Apparently a den may be dug long before the time of retirement. On 22 July 1953 I came upon a freshly dug den on Cathedral Mountain. The entrance was about 27 inches high and 24 inches wide. The tunnel was about 12 feet long and slanted upward slightly. At that time no chamber existed. On 23 August a chamber 4 feet by 3 feet had been hollowed out at the far end, the longer dimension at right angles to the tunnel. The den was intact 6 years later. I noted that cinquefoil bushes near the den had been nipped off in past years and brought into the chamber. Remnants of dry grass and herbaceous material also were present. One-quarter

Fig. 30. Dens used by bears for overwintering appear to be located throughout the summer range in the park.

mile from this den another den had been freshly dug that penetrated about 5 feet.

Another den on Igloo Mountain was dug on a gentle slope in a patch of willows. The burrow was 9 feet long. About two bushels of dry vegetation had been pawed out of the burrow and lay on the large dirt mound. The nest material consisted chiefly of blueberry, cinquefoil, and willows. Twigs were mixed with a lot of moss, and bear hairs were mixed in with nest debris.

The burrow of one old den had caved in but the roof over the chamber was intact. The chamber measured about 3½ feet by 5 feet and the burrow leading to it was about 9 feet long. Lambs were scampering about the den, and sheep had rubbed against the exposed sod.

In November 1920, O. J. Murie visited two unoccupied dens in the Savage River area similar to those I have described. One of these dens was dug in gravelly soil at the edge of timber near the base of a spruce where it had been necessary to bite off a number of roots. Because bears can so easily dig a den in loamy soil, it is probable that they dig new dens rather than search for one used previously. The 12 dens that I know about have been dug by bears. If a natural cave were available, I expect it would be used at times. O.J. Murie found a cave on the Alaska Peninsula occupied by a brown bear, and I was told that a natural cave at the head of one of the rivers in McKinley Park had been occupied. Black bears in more southern climes, such as Pennsylvania, may hibernate in a hollow, but I suspect that in northern country the bears seek the shelter of a burrow or a cave.

The dens I have visited were located throughout the bears' range so it appears that bears do not necessarily move into lower country to hibernate (Fig. 30).

Food Habits

The grizzly is a carnivore that cannot capture enough prey for subsistence. He hunts methodically mice and ground squirrels, but the small size of these rodents makes this hunting too time-consuming to satisfy his hunger or nourish his huge bulk. He is too slow to capture caribou, moose, or mountain sheep except for offspring a day or two old. On the coast of Alaska spawning salmon in some streams are an important food during some periods, but this food item is not available in the park. Carrion flesh is appreciated but occurs only sporadically.

To subsist, the grizzly has turned to vegetation for a staple, dependable diet. He has learned to exploit a variety of these foods.

I have summarized my 19-year observations of bear feeding in Table 7. This table demonstrates seasonal changes of foods and their relative contribution to the grizzly's diet. June is divided into two parts because a major change in food habits usually occurs in that month. Most of

Table 7. Tabulation of foods eaten by grizzlies in Mt. McKinley National Park from observations of feeding, 1945–1970.

Food item	May 1–31	June 1–15	June 16–30	July 1–31	August 1–31	September 1–30	Total
Roots	84	87	8	3	26	31	239
Total grazing	4	45	61	108	53	6	277
Unidentified	1	9	12	2	3	1	28
Grass	1	27	20	35	19	2	104
Horsetail	1	7	6	18	8	1	41
Willow	1	1	1	1			4
Oxyria			3	9	4		16
Rumex				4	1		5
Boykinia			1	22	17	1	41
Oxytropis			18	15	1		34
Coltsfoot		1				1	2
Sanguisorba				2			2
Total berries	9	11	11	11	61	10	113
Unidentified	2	1	7	3	19	1	33
Blueberry				6	16	2	24
Crowberry	5	9	3	2	13		32
Cranberry	2	1	1		2		6
Buffaloberry					11	7	18
Ground squirrel	5	2	4	10	21	15	57
Mice			1	3	2		6
Carrion	9	4	5	2	5	4	29
Total							721

these observations are from several years in the 1960s when I made a special effort to document food habits. In earlier years I did not always note what food was being eaten when I observed bears. Each observation is of a lone bear or a family unit. Details of food use are in the annotated list of grizzly foods and some of the sections on relationships with other animals.

Annotated List of Grizzly Foods in McKinley National Park

Roots: The principal food of the grizzly in the spring is the thick, fleshy root of the peavine (*Hedysarum alpinum americanum*). These roots become an important food again in autumn. The root resembles that of dandelions, and the flavor suggests garden peas.

Roots other than peavine also have been reported to be part of the grizzly's diet. In some diggings I have noted the exposed roots of rock fireweed (*Epilobium latifolium*) and possibly some had been eaten. The underground stems of coltsfoot (*Petasites frigidus*) appeared to have been eaten a few times. I have seen 7 or 8 feet of shallow sod composed of mountain avens (*Dryas octopetala*) rolled up like a carpet. But this may have been exploratory rather than to feed on the roots. One autumn I watched a grizzly digging roots near the top of a ridge on a steep, barren slope. From a distance I could detect only scattered cinquefoil (herbaceous) and a little rock fireweed. The bear moved about searching for plants with his nose, and digging a foot or more before reaching the root he sought. I thought that perhaps he was feeding on a root I had not recorded because peavine roots usually are near the surface and do not require much digging. The following day I examined the diggings and learned that the bear had been hunting peavine root. Because of the gravelly nature of the soil, the stems were long and the roots buried deeply. In practically all diggings that I examined the peavine was present and obviously was the species of root sought.

Root digging is the chief occupation of grizzlies during May and early June. In 1947 my latest record in spring for this activity was 10 June. In 1962, when there was an unusually heavy winter snowfall, and in 1963 when spring was very late, root digging continued undiminished until the middle of June, and the latest root digging noted in each of these 2 years was 21 June. The duration of the spring root-feeding period thus depends on the weather and the location (elevation, etc.) and it varies with the individual bear.

In late summer and early autumn some root digging is resumed, but berries usually continue to dominate the diet of most bears at this time. On 29 July 1953 I watched a grizzly dig a few roots; in 1961 a bear was digging roots on the East Fork bar on 7 August, and a few fresh diggings were seen on the Toklat bar on the same day. In 1960 two 2-year-old bears were observed digging roots almost daily from 8 to 25 September. In September there is considerable root digging, even when berries are available. At the time the two young bears mentioned above were occupied with digging roots, many other bears were feeding chiefly on berries. In 1963, a year in which the berry crop failed, more fall root digging than I had ever observed occurred.

On many old river bars and ridge slopes the peavine is abundant and distributed uniformly. Diggings often are so extensive and concentrated that they resemble plowed fields. One or both paws, usually both, are used to turn over chunks of sod and expose roots. When the paws are placed on the sod, the bear loosens a chunk with a series of pulling jerks, using his whole body in the effort. The roots exposed in the turned-over sod are then eaten, and more are uncovered by raking the soil from them with slow, delicate strokes. When small cubs are present, they may forage in the mother's diggings and uncover roots that she missed. When a bear starts chewing on a root, 6 or 7 inches of it may protrude from the mouth. A few times I have seen a bear use a paw to scrape dirt off a root held in the mouth. In places where sod has not yet formed and peavine plants are scattered, bears may move about more as they search for the plants with their nose. Like much of the vegetable food the bear eats, many roots seem to have undergone relatively little digestion when they appear in the scats (Figs. 31, 32).

Extensive areas over some of the old river bars have been rooted annually for years. In the digging, enough sprouts may develop to insure a continuous source of plants for the future. But in some favorite, more limited areas large roots apparently become depleted sufficiently to cause their neglect for a year or longer, until the young plants develop roots large enough to be attractive.

On the upper East Fork River, an old bar is covered chiefly with mountain avens. For 15 years or more it showed scarcely any digging by bears, but recently this extensive bar has been dug heavily, in both spring and fall. The plant succession was not monitored, but judging by the appearance of the present vegetation, the peavine has invaded the dense sod of mountain avens. In parts of the bar, adjacent to the diggings and stretching far beyond, peavine appears to be invading but the plants are still too young to have developed large roots.

Overflow ice, sometimes 10 feet or more in thickness, may form on some river bars during the winter and protect the peavine from the bears during the spring. If peavine is not available in spring, it would be during the autumn rooting period. These overflow ice deposits may vary in depth, extent, and specific location from year to year. Thus, a tendency

Fig. 31. Grizzlies expose succulent roots by loosening chunks of sod with their forepaws.

Fig. 32. Here grizzlies had been digging for the roots of peavine *(Hedysarum alpinum americanum)* in spring, far up the East Fork River.

for a natural rotation of rooting areas exists but may seldom develop enough to have a significant effect (Fig. 33).

When bears excavate roots on a slope the possibility of areas enlarging and developing progressive erosion is real. Generally, however, although the location of diggings can be noted in later years, a healing process sets in and the bare spots recover gradually. The disturbed but uneaten roots of peavine and other species in the diggings may sprout and form a good growth the same year the diggings occurred. The open slope above the east end of the Toklat bridge was excavated heavily by bears for 2 or 3 years in the early 1960s. In 1963 when I examined the slope after this rather heavy use, I found that the diggings had healed so rapidly that at a short distance they were not obvious. In 1964 and 1965 I saw no bears digging on these slopes.

The recovery on river bars also may be rapid. In 1962 I photographed fresh diggings on the East Fork River bar that were so contiguous that the area had the appearance of a plowed field. When visited the following year, with some expectation of taking additional pictures, most diggings were hidden by a new growth, especially of peavine which had sprouted from pieces of roots left in the turned-over sod.

Fig. 33. Winter deposits of overflow ice, sometimes 10 feet or more thick, give some bars a respite from bear-digging, since they do not melt until summer.

A small area along the Toklat River that was excavated thoroughly by bears in 1939–41, was, by 1963, grown over by dwarf birch, and a patch of cottonwoods 10 or 15 feet tall had grown up in one area that had been used heavily.

The braided channels of the rivers in the park are always shifting, invading and eroding old river bars covered with vegetation, and elsewhere permitting, over the years, the development of new vegetated bars. Thus these areas, used by bears as a source of roots, are not static. In a national park our policies protect these natural processes so that no effort is made to freeze the environment at some particular stage.

In other parts of Alaska roots also are an important part of the grizzly's diet. O. J. Murie examined 151 scats gathered in the upper Sheenjek River area. The scats were not dated accurately, many of them being old when found, but they showed that bears fed extensively on peavine roots. Fifty-five of the 151 scats collected contained peavine roots. O. J. Murie (1959) stated that spring food for the brown bear on the Alaska Peninsula consisted chiefly of grass and roots. On Montague Island, Sheldon (1912) reported brown bears feeding on the roots of skunk cabbage *(Lysichiton).*

In the scat table for McKinley National Park (Table 8), note that 105 of the 810 bear scats examined contained roots. Of these 105 scats, 82 contained only roots. These figures and my observations indicate that when bears feed on roots they concentrate on them almost exclusively.

Grasses and Sedges: The spring root diet is abandoned as the new green vegetation becomes available. This may be in late May or, in the higher elevations (3,000 to 4,000 feet), during the first 2 weeks of June. In 2 years when the season was late, green grass was not eaten until 15 June, and in one year, not until 18 June. In those years, feeding on roots continued longer than usual. There may be a considerable overlap between the spring root-feeding and grass-eating periods. Bears may still be feeding extensively on roots when they first begin to find patches of green grass. During this transition period, some bears may be feeding only on roots when others have discovered that green grass is available and are concentrating on it.

The first grass available is a tall species (*Calamagrostis canadensis*). It is not a favorite but because it appears early, it is sought eagerly. To get at the green shoots the old mass of dry stems and blades sometimes is pushed aside with muzzle or paw. At this time of grass scarcity a bear may make a full bite to get only a single grass shoot—a ludicrously big effort to get so little. When vegetation growth begins it is rapid and favorite green foods become so available that early spring grazing on this particular grass soon ceases.

The favorite grass is *Arctagrostis latifolium.* This species resembles *Calamagrostis* but bears have no difficulty differentiating between them.

Table 8. Occurrence of food items in 810 grizzly bear scats collected in Mt. McKinley National Park, 1947-1970. Numbers in parentheses are occurrences of 50 percent or more of the item in scats. The number of scats examined in each period is shown at the top of each column.

	May 1-31 37	June 1-15 60	June 16-30 96	July 1-31 144	Aug. 1-31 285	Sept. 1-30 177	Oct. 1-15 11	Total 810
Roots	20(19)	19(16)	4(3)		7(5)	49(45)	6(6)	105(94)
Grass	3(1)	25(21)	59(48)	80(66)	69(50)	25(17)		263(203)
Herbs	2(1)	2	30(8)	66(22)	54(27)	19(11)		173(59)
Horsetail		13(13)	49(44)	8(6)	25(16)	2(1)		97(80)
Oxytropis			7(2)	67(67)	66(62)			140(131)
Boykinia					4(2)	1(1)		5(3)
Blueberry	1(1)	2		1(1)	65(26)	72(51)	2(2)	143(81)
Crowberry	6(5)	5(3)	5(3)	2(1)	136(111)	94(61)	6(4)	254(188)
Buffaloberry				2(2)	29(17)	55(31)	1(1)	87(61)
Cranberry	6(1)	3(2)	7(1)			3		19(4)
Ground squirrel	3	3(1)	2	2	5	15(1)		30(2)
Marmot		1(1)	1		1(1)			3(2)
Mouse								0
Caribou (adult)	2	2(1)			2(1)	3(3)		9(5)
Caribou calf	5(3)	5(1)			2(2)			12(6)
Mountain sheep	4(4)	2(2)						6(6)
Bear				1				1
Ptarmigan		1						1
Wasp	1				1			2
Willow twigs				1				1
Total number of occurrences								1351(925)

The juicy-stemmed *Arctagrostis* grows in moist hollows and draws and along streamlets. It is associated closely with palatable herbaceous species, so that in his grazing the bear may feed for a time on grass and then shift to some of the herbs. *Arctagrostis* is perhaps the most important of the green foods in the areas I observed in the park.

When berries become available later in July and early August, the feeding on grass and other green foods slackens and the bears turn to these fruits. In areas where the berry crop is good the grass feeding may be abandoned rather abruptly.

In years when the berry crop is generally poor, grass continues to be eaten throughout August and in early September. In 1963, when the season was late and the park suffered an almost complete berry failure, bears continued to feed extensively on green vegetation during August and well into September. On 27 August 1963 two families on Sable Pass fed throughout the day on grass, sedges, and herbs. Very little sedge is eaten as a rule, but at this time, because the snowfields in the hollows were slow in melting, sedges were still young and tender whereas most other green foods were old and tough. On 11 September 1963 a family and a lone bear fed extensively on green foods including a sedge (*Carex podocarpa*). On 16 September a fresh scat in the Thorofare area contained only blades of mature sedge that were still green but no longer tender in a wet, flat area nearby. In 1964 the berry crop was also poor and as a result green foods were eaten extensively during the usual berry-feeding season.

Bears swallow grass and sedge with relatively little mastication and much of it appears in the scats, little altered after its passage through the digestive tract.

Grizzlies have been found feeding on grass and sedge in other areas. On the Sheenjek River (Brooks Range) O. J. Murie recorded grass and sedge in 46 of 151 scats. He found the brown bear feeding on grass on the Alaska Peninsula in early June. Sheldon (1912) reported brown bears on Montague Island feeding on a special kind of grass above timber. In Yellowstone National Park I have seen grizzlies grazing on grass as early as 11 May. In Glacier Bay National Monument I saw a black bear grazing steadily in a large patch of sedges, and it is probable that grizzlies there would seek the same sedges.

Of the 810 scats examined, grass was found in 263 of them. Seventy-nine of the scats contained only grass.

Horsetail: Horsetail (*Equisetum arvense*) is relished by both grizzlies and black bears. Feeding on horsetail begins in late May, as soon as the new green growth becomes available, and continues through the summer. As late as 28 August I have found fresh droppings containing only horsetail. On 6 September 1964 I saw a mother and yearling feed on a fresh growth of horsetail. (At this time most of the plants were old, and some patches had turned brown.)

In looking through some of O. J. Murie's notes I found a few references which showed that black bears in Alaska are also fond of horsetail. Horsetail also is relished by Dall Sheep and is eaten by ground squirrels and Willow Ptarmigan.

I have watched grizzlies feed steadily in a patch of horsetail, then switch to sourdock, *Boykinia*, and *Arctagrostis* (grass) for a change. Horsetail is one of the grizzlies' favorite summer foods. It is possible that some of the other species of *Equisetum* are also eaten occasionally but I have no such records, and one or two species growing in ponds and eaten by moose probably are palatable to bears.

That bears concentrate frequently on horsetail is indicated by the fact that of the 97 scats that contained this plant, horsetail made up 100% of 53 of them.

*Saxifrage (*Boykinia richardsonii*):* The showy *Boykinia*, with its large rounded leaves and conspicuous cluster of white blossoms growing in a spike, occurs in damp hollows, along streamlets, and in openings among tall willows. It is often associated with other food species eaten by grizzlies. Flower heads, leaves, and parts of stems are consumed. A bear may at times take a liking to the flowers and move along cropping off one flower cluster after another and neglecting the leaves. Once, however, I watched a bear bite off stems 6 to 8 inches below flower heads, eat the stem and leaves attached and discard the flower. Usually, a bear feeds briefly on this plant, then turns to grass, horsetail, or other species if they are associated, shifting back and forth between species. Once I watched a bear in an extensive patch of this big-leaved plant, concentrating on it for 2 hours. In 1963, when berries were scarce, I saw much late summer feeding on *Boykinia*. On 27 August some of the plants whose growth had been delayed because of late melting snowbanks were sought eagerly. The stunted, bunchy plants were grazed almost completely, the bears biting off one leaf at a time, doing a thorough job rather than picking haphazardly here and there as they usually did. In that same year the entire contents of a fresh scat examined on 22 September contained only *Boykinia*, an unusually late record. This species, along with some of the other herbaceous food plants that generally are not listed specifically in scat tables, is underrepresented in the scat analyses. Many of the droppings containing grass also had herbaceous remains.

*Sourdock (*Rumex arcticus*):* Sourdock grows luxuriantly in moist hollows and along small streams and is eaten extensively by bears. On 7 July 1960 two 2-year-old cubs fed on this species for 30 minutes, although other favorite foods such as horsetail, *Boykinia*, and grass were present. After a siesta, the two cubs fed for a long period on grass (*Arctagrostis*), then for a time on sourdock again, and later on horsetail. Only the leaves and stalks of the sourdock are eaten. The seed stem is bitten off 6 or 7 inches below the large seed head and maneuvered into the mouth so that the seed head is cut off and discarded. On 22 July 1961 I watched two spring cubs feeding on sourdock. They bit off the stem near the ground, ate stem and leaves, and discarded the seed heads as deftly as did older bears. While I watched, the mother of the cubs did not feed on this sourdock but sought other species of vegetation.

*Mountain Sorrel (*Oxyria digyna*):* One of the first sources of green food in early summer is mountain sorrel, with its round leaves and sour taste. In 1963 a mother and two 2-year-old cubs were observed grazing on the mats of this plant that grew in a draw among tall willow brush. The growth was so short that the bears practically had to gnaw it off the ground, yet they fed extensively on it. This species is eaten frequently, but is a less important item than some other herbaceous food plants because of its limited availability.

*Viscid Oxytrope (*Oxytropis viscida*):* This species of peavine grows extensively on old river bars, especially near the headwaters. In late June and much of July some bears spend hours grazing on the flowers and leaves. One year in June, at the head of East Fork River, I saw seven bears (a mother and two spring cubs, and two sets of twins about 3 or 4 years old) feeding on this species. In 1965, from 7 to 12 July, two families fed extensively on this species on the west branch of East Fork River. By August most grazing on this plant had terminated. Another year in August, I found several old scats at the head of East Fork River that contained this species, but in fresh scats only a trace was noted in one of them. Although some bears feed a great deal on this species, others apparently seldom visit areas where it is plentiful and consequently use little of it. Viscid oxytrope only was present in 131 of the 140 scats in which the species occurred.

*Willow (*Salix spp.*):* I have noticed grizzlies eating willow on only a few occasions. On 4 May 1940 a yearling cub was observed eating a few catkins, and on 4 June 1955 another yearling cub was seen feeding briefly on them. On 23 May 1961 I watched a bear biting casually at willow twigs as he walked steadily on his way. On 23 June 1962 a mother and yearling ate a few willow leaves.

Bergman (1936) found that willow catkins are an important early spring food of the Asian grizzly in Kamchatka. He writes: "Before the hills become green, willow catkins are eaten with the greatest relish. Hunters agree that these catkins play a great role in the springtime food of the bear; one frequently sees willow bushes, stripped of their catkins, surrounded by bear tracks."

It is possible that the McKinley bears feed more on catkins than my observations indicate. Bergman makes no mention of grizzlies feeding on roots, so perhaps in McKinley catkin feeding is replaced largely by rooting. However, it would not be surprising to find bears feeding extensively on catkins in early spring in years of heavy snowfall.

Mushrooms: No mushrooms were found eaten in McKinley National Park by grizzlies, but, they are plentiful and probably are eaten occasionally. O. J. Murie found mushrooms in 8 of 42 scats he examined in Yellowstone National Park. The percentage present varied from a trace to 100%. Mrs. Ruth Onthank wrote me that she often had seen coral mushroom (*Clavaria*) dug out before it could break through to the surface, and signs indicated that black bears had been feeding on them. In France, of course, pigs feed extensively on the subterranean truffle.

Spruce Cones: I have found spruce cone remains in only one dropping in the park (Murie 1944). In Yellowstone and Teton National Parks the cones of whitebark pine (*Pinus albicaulis*) are eaten by grizzlies and black bears. The nuts in the cones of spruce available in McKinley National Park probably are too small to be sought after. If the cones were palatable, they would be abundant when there is a good cone crop, because in those years red squirrels collect large caches of them on top of the ground in autumn.

*Blueberry (*Vaccinium uliginosum*):* Grizzlies begin feeding on blueberries before the fruits are fully ripe. At the higher elevations, from 2,000 to 4,000 feet, feeding usually begins in the last half of July. The earliest record I have is 12 July, when blueberry was the principal component of a fresh dropping. Bears bite at the low bushes in much the same manner as they graze on grass, stripping leaves and berries. Occasionally, a paw is used to raise a heavily laden branch to bite more easily at the fruit.

Blueberry bushes, growing a foot or two tall, are abundant and distributed widely. The crop varies from year to year, but at some lower elevations the berry crop seems to be uniformly good. The bears seek out the best patches and for hours on end bite vigorously and rapidly at the bushes.

I have witnessed only a few poor berry crops in McKinley National Park. In 1963 all species of berries eaten by bears were scarce in most localities and not abundant anywhere. Various reports indicate that the berry crop was substandard in much of Alaska that year. Lateness of spring may have accounted for this crop failure. In 1964 and 1965 the overall berry crop was again below standard; poor in higher elevations and plentiful only in spots at lower elevations. Crowberries were plentiful, however, and bears fed heavily on them. Of the 143 scats in which blueberry was found, 31 contained only blueberry.

*Crowberry (*Empetrum nigrum*):* Crowberry is distributed widely over the park, growing in the woods and open country and far up the slopes. The crop is usually excellent and bears eat great quantities of it. This species vies with blueberries and buffaloberry in popularity, and because the berries winter well, it supplements the spring diet of roots. I often have watched bears feeding on crowberries in May and June. Some pass through the bear's digestive tract unbroken and others are only crushed. Of the 254 scats which contained crowberry, 138 had only crowberry.

*Buffaloberry (*Shepherdia canadensis*):* Buffaloberry grows on rather gravelly terrain. It is found on lower slopes, washes, and old river bars. The berries usually are abundant and an important source of food. The berries become available during the latter half of July; I have seen bears feeding on them as early as 16 July. Many leaves are eaten with the fruit, no doubt inadvertently. At times mountain sheep compete for some of the berries but not seriously. In 1961, during a September migration, about 130 sheep stopped as they were crossing the river from Divide Mountain and fed on a bar along the Toklat River. A number of sheep fed on buffaloberry, depleting slightly the supply in this restricted area. Of the 87 scats that contained this species, 38 contained only buffaloberry.

*Silverberry (*Eleagnus commutata*):* Over most of the park silverberry is not available. I saw two small patches near Igloo Creek, and noted flowers on the bushes but no berries. Down the Toklat River, near the park boundary, I have seen several bushes and it is likely that at these lower elevations the plants bear fruit. The mealy berries remain on the bushes all winter. Coyotes in Grand Teton National Park sometimes feed on them in winter.

*Cranberry (*Vaccinium vitis-idaea*):* Cranberry is distributed widely and the plants bear heavily. The berries are not eaten much in late summer or autumn, but the grizzlies consume some in the following spring, during May and early June. On 20 June 1955 fresh scat from a yearling contained chiefly cranberry, and on 20 May 1962 a fresh scat left by a large male contained mostly the hair of mountain sheep but also about 600 cranberries. In 1962 a few cranberries were seen in fresh droppings as late as 21 and 25 June.

Arctostaphylus alpina *and* Arctostaphylus rubra: The large, black berries of *A. alpina* that grow in the open country, and the red berries of *A. rubra* that grow in moist areas and in woods are eaten occasionally. However, they are so scattered on the plants that they are not eaten in quantity in the park. On 7 August 1926, O. J. Murie found a number of the red berries in the stomach of a black bear killed on Old Crow River, and in the same locality berries were found in many droppings of the black bear.

*Rose (*Rosa acicularis*):* I have not witnessed grizzlies eating rose hips, but in low country where the rose is plentiful it probably does enter the diet. Along the Porcupine River, O. J. Murie found that black bears eat rose berries, and I have seen them eaten in Yellowstone National Park by black bears. In McKinley National Park the rose would not be an important food item.

Miscellaneous Foods: Various other plant foods are tasted occasionally but are unimportant. I have seen *Artemisia arctica, Sanguisorba sitchense, Polemonium* sp., *Heracleum* sp., and *Angelica* sp. tasted. On one occasion it appeared that the underground stems and buds of coltsfoot (*Petasites*) had been eaten.

Figures 34–39 depict various plants used for food by bears.

Fig. 34. The fleshy roots of peavine *(Hedysarum alpinum)* are a principal food of bears in the spring.

Fig. 35. The juicy-stemmed *Arctagrostis* is the grass species eaten most frequently by grizzlies.

Fig. 36. *Boykinia,* a showy saxifrage, is a major food item in favored grazing areas on Sable Pass.

Fig. 37. One of the attractions for bears in the lush, moist grazing areas on Sable Pass is the sourdock *(Rumex arcticus).*

Fig. 38. Some bears spend hours grazing on *Oxytropis viscida*, a member of the pea family.

Fig. 39. Blueberries *(Vaccinium uliginosum)* are usually abundant in the park and are a major bear food in late summer.

Caribou: At all times these animals are a potential source of carrion (discussed under "Carrion"). During the calving season some calves are captured by bears that live in calving areas. (The mammal species that are food for grizzlies are listed here, but also are discussed in detail in the sections dealing with grizzly relationships and with carrion.)

Moose: Adult moose furnish a certain amount of carrion, and a few calves are obtained in the calving season.

Mountain Sheep: Sheep eaten by grizzlies usually is carrion. It is possible that a sheep may be captured occasionally when migrating across stretches of flat country, although I have no record of such an occurrence. Occasionally a bear has been seen chasing migrating sheep in the Toklat River area, but the sheep escaped. Once a migrating ewe was in danger when crossing a late spring snowfield, but she managed to cross without being overtaken by the pursuing grizzly. On one occasion a yearling was captured. Sheep are a sporadic source of food.

Shed Moose Antler: On 8 August an employee of the National Park Service saw a large bull moose drop an antler on Sable Pass. Later, I saw a bull with only one antler; it was an animal that had spent the summer in the area. When I found the antler, a few days after it was dropped, a bear had eaten the soft tips and the velvet. The antler had been infected and necrosed just above the pedicel, causing it to drop off. On two occasions I have seen cubs chewing on a caribou antler, but I think this was done in the spirit of play.

Marmot: The marmot apparently is seldom captured by the grizzly. A marmot den usually is located in rocks or cliffs where bears cannot dig it out. It would seem that if marmots dug dens away from rocks, bears would capture them more frequently.

Ground squirrel: The ground squirrel makes up a small but perhaps important part of the grizzly diet. It is eaten at all seasons, and may be hunted methodically.

Voles and Lemmings: Meadow mice and lemmings furnish the bears with a taste of meat. They are not particularly sought after, but when they are abundant, bears may spend some time feeding on them.

Beaver: Bears probably seldom capture beaver but on occasion they may discover one that is too far from a pond to escape. On 30 May 1941 a female beaver containing two large embryos was found dead on the shore of a creek near Wonder Lake. It was potential carrion.

One year in late June, Mrs. Elizabeth Berry saw a bear working at something on the shore of Wonder Lake. When the bear left, a beaver carcass was discovered frozen into the ice that filled a burrow. The grizzly returned later and retrieved the carcass.

Grizzly Carrion: I was unable to determine the extent to which grizzlies will feed on grizzly carrion, and possibly it varies with the individual. I have driven sled dogs that exhibited individual variation in their tastes for dog meat. Some ate it raw, whereas others, although hungry, would not eat it uncooked. A female grizzly that killed two spring cubs did not feed on the carcasses, although, at the time, she apparently was too concerned over the safety of her own cubs to be interested in food. However, I have seen two carcasses that were fed upon by grizzlies.

Insects: In McKinley National Park few insects are available for food. A few wasps are eaten; about a dozen were found in one scat and two or three in another. Apparently the ground nests had been dug out. O. J. Murie found two bees in a scat in the Sheenjek River area and some in a scat collected in Yellowstone National Park.

On some grizzly ranges farther south, insects are more important in the diet. O. J. Murie found ants in 10 of 42 scats collected in Yellowstone National Park.

In the Mission Range, Montana, a number of bears were observed turning over rocks above timber, at an elevation of about 10,000 feet (Chapman et al. 1955). These authors found large aggregations of ladybird beetles (*Coccinella*) under rocks in the area where bears were feeding. Some years later, one of the authors collected and examined 15 grizzly scats in the area at an elevation of 8,000 to 9,000 feet. Nine of the droppings consisted almost entirely of moth (*Chorizagrostid auxiliaris*) remains, the adult stage of the army cutworm. Ladybird beetles were not found in the area at the time the droppings were collected. The authors stated that the moths apparently were captured under rocks where they gathered during the summer.

In Yellowstone National Park I noted black bears on the summer buffalo range feeding on grasshoppers and Mormon crickets. The bears had turned over hundreds of buffalo chips to find these insects. Of 64 scats collected, 61 contained Mormon crickets and grasshoppers, mostly the former (Murie 1937). If grizzlies had been present, no doubt they also would have fed on these insects.

Data from Scat Examinations

The food habits of grizzlies in McKinley National Park can be determined satisfactorily by watching bears feeding and checking feeding signs. Additional data were secured by examining scats and estimating the approximate proportions of various food items. Since the scats usually could be dated fairly accurately, the data in Table 8 are segregated into time periods, indicating seasonal food habits. The number of scats in which various food items occur depends to a considerable extent on how much time was spent collecting in the different habitats, and some items, such as *Boykinia* and mice, probably are underrepresented in this analysis. Moose are not recorded in Table 8, yet we know that it was eaten. Moreover, I only identified some of the herbaceous material, and sedges were included with grass. The importance of various herbaceous species is cited in the annotated list of food species (Figs. 40, 41).

The proportion in which various food items were represented in the scats were calculated, but this classification, usually done in the field, was quite rough except for those scats that contained only a single item. I have indicated in Table 8 (numbers in parentheses) the number of occurrences in each category that represented 50 to 100 percent of the scat.

Figs. 40,41. Additional information on food habits can be gained by examining bear scats; a scat containing mainly horsetail (above) and one composed of crowberries (right page).

Summary

During spring, the chief food is the root of the peavine, and bears can be seen seeking these roots on old river bars and ridge slopes, rooting areas well known to them from previous years. This root diet is supplemented occasionally with crowberry and cranberry, berries that survive the winter fairly well; blueberries also show up in a few spring scats. For those bears living on migration routes where caribou calves are born, the calves may be an important food item in part of May and early June. A few moose calves also are secured at these times. When green vegetation is available, the diet of spring roots is dropped.

Green grass may be available in late May or early June, but in some years, in places such as Sable Pass where the season is late, this food may not be available in quantity until the middle of June. This change in diet is sought eagerly, and bears feed avidly on the first green shoots.

Not all bears turn to green grass at the same time, however, even in the same general area; some continue feeding on roots after others have found and are feeding on grass. The first grass to appear in many places is a tall species, *Calamagrostis canadensis*. It apparently is relished when young and the only green food available, but when horsetail, certain herbaceous species, and their favorite grass, *Arctagrostis* , appear, the bears turn to them. The palatable grasses and herbs generally grow in moist hollows and draws, in hummocky areas, and along small streams. One of the heavily grazed legumes is abundant on old river bars. Grass, herbs, and horsetail, the staff of life during much of June and July, are attractive to bears until berries begin to ripen.

When berries (chiefly blueberry, crowberry, and buffaloberry) ripen, they are the favorite food. Green grasses and herbs are eaten after berries enter the diet but usually are second choice. Bears devote most of their time to berries, and a majority of the droppings contain only these fruits. Blueberry bushes are dispersed widely in extensive stands. They are found in valleys, on flats, and on lower ridge slopes. Another important berry, crowberry, also is dispersed widely over the landscape, and these evergreen, trailing plants are loaded with juicy, black berries. The bitter, red buffaloberry also is plentiful and usually produces a large crop. Its distribution is more localized than blueberry and crowberry, being usually found on old gravel bars, on dry benches near streams, and scattered on lower ridge slopes. These three species comprise the bulk of the berry diet. Cranberry is abundant but is not eaten extensively. *Arctostaphylus* berries are eaten but are not abundant enough on the plants to be much sought after, although I have found a few scats containing many of these berries. Currants, found commonly in alder thickets, probably are eaten.

In late August and September some bears return again to roots. In September, in poor berry years, I have seen some bears feeding on roots all day for several days while others persisted in seeking berries. In those years when the berry crop is poor, green grasses, sedges, and herbs may remain important through August and into early September. There are individual differences in food habits, especially during transition periods, due in part to the food supply where a particular bear is foraging. Some bears that summer at higher elevations regularly seek lower country when the berry season starts, but others remain where they are.

Ground squirrels are eaten at all seasons, but particularly in late summer and fall. At any season when bears are abroad one may chance to see a bear working industriously to dig out a ground squirrel, a task that varies in duration and may be unsuccessful. In years when fieldmice or lemmings are abundant, some bears may be seen feeding extensively on them.

Carrion always is attractive to meat-hungry bears and is available quite often. Thus, it can be seen that the diets of grizzly bears are varied.

Carrion and Caching

Carrion is attractive to many animals and especially to meat-loving grizzlies. The important sources of carrion in the park are caribou, moose, and sheep. Sooner or later individuals of these species succumb to disease, predation, or old age (directly or indirectly) and become sources of food. The magpie, camp robber, golden eagle, wolf, fox, lynx, wolverine, and grizzly all seek their full share. If the direction of the breeze is favorable, the scented message may reach the grizzly from afar, and he may be the first to reach a carcass. If he is not so fortunate, the carcass may already be devoured by the competition when he reaches it. Sometimes a carcass is not discovered for several days, but the state of decay is immaterial to a hungry grizzly.

The supply of carrion in a bear's home range varies from year to year, and these blessings are not distributed uniformly. Some grizzlies may be located favorably to procure carrion because of the prevalence of large herbivores on their range. For the most part carrion represents only a special treat.

In August 1962 several bears in country I was frequenting were especially fortunate in regard to carrion. These bears were living on the migration route of the caribou and, although the main caribou herds had migrated westward, a number of scattered caribou bands still remained. During this August I knew of six old caribou bulls that had died, and the bears probably knew of others. One old bull, apparently ailing, was killed by a lone wolf, but the bears did not recognize any special wolf rights and helped themselves. One-half mile away another old bull caribou carcass was untouched for a few days. This bull was in good flesh but it apparently had died from a disease. The bears, wolves, and others had devoured all evidence of the other four dead bulls by the time I examined them, but all had reached a ripe old age.

After eating to his capacity, a grizzly usually covers the carcass with sod and debris. He paws the debris loose first with one paw then the other, then rakes it back toward the carcass. Sometimes, after loosening and pawing debris with his forepaws, he may scrape it farther back with a few hind-foot strokes. It is a lazy process, undertaken after a huge meal. After the carcass is well covered, the raking may be resumed at intervals as though some satisfaction is connected with this activity. When the carcass is heaped over with scrapings, the bear may rest nearby, or he may lie on top of the cache as though proclaiming his proprietorship. Sometimes after covering a carcass the bear may move off a short distance to rest. If one comes upon a bear cache, the bear probably is resting nearby or has fled at one's approach. Even after only bones and hide are left, the carcass may be visited occasionally. Once a mother followed by two yearlings, after feeding, retired to cliffs one-quarter mile away from the carcass, but the next day after feeding she rested near the carcass.

A bear may roll on an old carcass just as a fox or dog will. One day I watched a 3-year-old bear investigating an old caribou carcass long since salvaged. Before leaving, the bear rolled on the remaining pieces of hide and bone for 3 minutes. Then he walked to a little creek, waded into a deep hole, and lay down so that the water covered him except for his head and shoulder hump. Bears often lie in water in this manner, apparently to cool off, so I doubt that he was bathing or trying to wash away odors.

Usually no contact occurs between a bear at a cache and an intruder. Superiority seems to be recognized. Either the bear at the cache will retreat or, if he is bigger or perhaps more aggressive, chase the intruder. Very likely the bears on a range are well acquainted and do not need to re-evaluate status at each meeting. Occasionally an altercation takes place, because an intruding bear, even if small, is attracted strongly by carrion and will make some effort to partake.

During the winter when bears have tucked themselves snugly into dens, much carrion becomes available. Most winter carrion is consumed by wolves, wolverines, and foxes. But when bears first emerge in early spring, they salvage some of the leavings. They may find a sheep skull they can crush to retrieve the brain or bits of nourishment on bones and hide. Sometimes winter kills are covered by snow and become available when thawing begins and bears are abroad again. In the winter of 1962–63 a number of caribou succumbed on a mountain slope near Deniki Lake, just east of the park. Apparently they were killed in an avalanche. Bears fed on the carcasses throughout much of the summer. At least six or seven bears, including cubs, were attracted to the slope and were observed over a period of several weeks by Bill Nancarrow and Mr. and Mrs. William Berry. As summer progressed, the snow melted and additional animals were uncovered.

The following incidents, a few of many I have observed involving caching behavior of grizzlies and other activities at carrion, show more fully the ways of bears with carrion.

A Mother With Three Cubs Concerned About Her Caches

On 24 August 1906, Sheldon (1930) describes a mother with three cubs that hurried to the carcass of one of seven rams Sheldon had shot and began:

> . . . to paw out rocks near the carcass, scooping out a deep hollow, tumbling big rocks down the canyon and moving others to one side, apparently with no effort at all. Then, seizing the carcass with her jaws she dragged it into the hollow and pawed rocks all around it, completely covering it, so that nothing but a mound of broken rock was visible.
>
> The bear then ascended the canyon to the carcass above and buried it in a similar manner, all three cubs now running around her, watching with curiosity. When she

started up the side of the canyon followed by two of the cubs, the other one, having discovered that the carcass was palatable, pawed off some of the rocks and remained behind, eating. When the old bear reached the top and looking back saw the cub, she rushed down and gave it a cuff, causing it to bawl loudly and run to one side. Then she pawed back the rocks and, followed closely by the three cubs, went to the top and started back across the slope.

She had not gone far when she suddenly turned, ran back to the edge of the canyon and gazed below as if to reassure herself that the carcasses were in no danger of being disturbed. Again she started across the slope, but again rushed back for another look. This she repeated fourteen times during twenty minutes, the last time running back at least a hundred yards.

After travelling somewhat over a quarter of a mile she and the cubs lay down among willows. Later in the afternoon she was seen feeding on a third ram carcass that she had cached on the slope. One of the sheep carcasses had been dragged across a steep slope three or four hundred yards before caching.

Bull Moose Becomes Carrion

On 16 November 1949 a bull moose, carrying a huge set of antlers, was resting on a willow patch near Savage River. Looking at him through field glasses, I noticed a whiteness on his eyes. When he stood up and browsed on willow, he was inept and groping and when he walked, each front foot was lifted high and moved forward uncertainly as though feeling for the ground.

There were scars about his face that may have been made by antler points, suggesting that the bull had been blinded in a fight during the rut. As he faced my companion and me, his ears were laid back as they are when a moose is angry; he licked his lips, and saliva drooled from his mouth.

This was a last stand, not against a predator, but against an infirmity. Until recently, he was a monarch whose antler spread was great and who could hold his own against a wolf pack or grizzly bears. It was sad to contemplate. Now he lived in a world of darkness, feeling his way and stumbling against boulders. I wondered if wolves would notice his condition and how his head jerked when an antler struck unexpectedly against a spruce or willow limb. It would have been an act of mercy to kill him. However, he was in the middle of an extensive area of willows and did not lack nourishment.

The following day we returned to the bull to take pictures. He was lying down about 400 yards from where we had left him. During the interim he had lost an antler, which we found. The one he still carried had an abnormally located prong, and we found it later after it had been shed. One antler weighed 25 pounds, the other 24 pounds; we estimated that the antler spread was at least 70 inches.

The bull was still alive on 19 January, but on 15 February we found his carcass and judged he had been dead 3 or 4 days. The temperature was about −20°F and the carcass was frozen. In time, successive snowfalls buried it. On 28 February the carcass was visited by a wolverine, a fox, and ravens which had fed on the neck. During the next month and a half, the above three species and a black wolf visited the carcass regularly. If the carcass had been fresh, it would have disappeared in a matter of days, but being frozen solidly and covered with packed snow, the flesh could be removed only by laborious gnawing and no parts could be carried away for caching. Enough remnants still were available after the middle of April to interest a grizzly that was abroad early and wandering over the country looking for food.

That same spring, on 11 May, I saw another large bull moose that had survived the winter but would soon become carrion. He was extremely thin, and his poor condition was further evident from the fact that no new growth of antlers was noticeable. In healthy bulls at this time new antlers had attained a length of 10 or 12 inches.

Lone Bear Discovers a Carcass

About 9 a.m. on a warm day in July, I watched a lone grizzly with nose raised, ears cocked forward, traveling into the wind along a hillside on a contour. He stopped a few times to look ahead. He was approaching the carcass of an old caribou bull. The bear's behavior suggested that he was about to make a discovery. One hundred yards from the carcass he stopped to look and test the breeze. From there he moved more eagerly, and nearer to the carcass he found a leg bone with most of the meat removed. After feeding on it for a few minutes, he walked down a bank, which lay along a little stream, to the carcass and managed to pull it part way up the bank. He started feeding on the rib section, swallowing large pieces but chewing little. In half an hour he moved off a few yards and scratched himself thoroughly with front and hind paws. He resumed his feeding for a time, then drank briefly at the stream. He rested for 15 minutes, then began pawing the sod, sending the chunks in the general direction of the carcass. After a few minutes, he lay down with ears cocked and at short intervals raised his head to look around.

At 10:30 a.m. he resumed feeding, and in 40 minutes was again scraping sod toward the carcass. At 11:20 a.m. he had another drink, then alternated resting and pawing sod and debris toward and over the carcass from 12 or more feet away. Once he trotted over to chase away two magpies. At 12:40 p.m. and again at 1:20 p.m. he took another drink. At 3:30 p.m. he uncovered part of the carcass and ate. By 4:15 p.m. he had pulled the carcass farther up the bank, and a little later he started covering it with the debris he had pawed loose earlier.

When I saw the scene the next morning, the carcass was well covered and only the antlers protruded. The bear was engaged part of the time in loosening more sod and pawing it over the carcass.

In the afternoon a slightly smaller bear intruded and was charged and chased for some distance. Later a second bear was reported to have crossed the stream three times, each time making contact with the bear in possession and each time being repulsed.

Bears Feed on Two Sheep

On the afternoon of 22 June 1965, a dark mother and a 2-year-old cub were discovered as they fed on a carcass lying in a short ravine not far from Toklat River. The carcass had been covered with debris by the bears, and was partially uncovered where the bears fed. Not enough of the carcass showed for identification. After feeding to a state of satiety, the mother pawed more debris over the cache, then collapsed on the heap as though too full to stand. A little later the cub, who had been feeding at one edge of the heap, approached the mother, pushed his head under her neck and chest and succeeded in promoting a nursing.

The following morning (23 June) the family was at the carcass when I arrived at 7 a.m. They fed, rested, and nursed until 4 p.m., when they moved into the willow brush and fed for an hour on horsetail before climbing a slope to rest. As they lay on the hillside some 150 or 200 yards from the carcass, a blond mother followed by a blond yearling emerged from the heavy willow brush and gleaned what they could from the remains. Two hours later they moved off and lay down 50 yards away.

On the morning of 24 June a magpie was foraging on the cache, finding crumbs too small to be noticed by bears. Fifteen minutes later the dark mother and her 2-year-old cub emerged from the willows for another taste. The cub jerked at a part hidden from its view and exposed the skull of a ram with large horns. Later, when I examined this head, I learned that the carcass was that of a ten-year-old ram. Apparently, this old, vulnerable animal had wandered onto gentle terrain in search of green vegetation, been surprised while away from friendly cliffs, and been killed by wolves whose tracks were at the carcass.

One-half hour after the family arrived at the carcass, a large male bear came up a small creek bed and turned into the short ravine where the family was feeding. The frightened family departed and the male took over. He found little attraction at the cache and after stirring around for 15 minutes, while a pair of shrikes snapped their beaks to demonstrate their annoyance, he returned to the brushy hillside. Later, I caught a glimpse of the blond mother and yearling in the tall willows near the cache, but they moved away to continue grazing on horsetail.

Before arriving at the carcass of the old ram on 23 June, I had discovered another carcass of a 4-year-old ram one-quarter mile or less to the south. Tracks crossing the road above this carcass showed deep tracks made by the ram, and deep imprints of a wolf. Apparently, a wolf had chased the ram and overtaken it a few yards below the road. The carcass was only 60 yards down the slope and had been dragged part of this distance. Most of it had been devoured. Earlier that morning, three wolves had been reported near the carcass. When the stomach contents were examined, I remarked to my companion that the food was masticated poorly. Many leaves of willows, including *Salix reticulata*, and leaves of anemone were intact; there were even clusters of willow leaves. This suggested that the ram had difficulty chewing, and when I examined the teeth, I found them badly ulcerated and worn on a slant. One molar had dropped out and another was loose in its ulcerated socket. On one side of the lower jaw an ulceration in the bone contained about a tablespoonful of pus. This was another classic example of a weakened sheep, in this case from disease, being eliminated by a predator.

The carcass of this 4-year-old ram was visited briefly by three wolves on 25 June but so far as I knew was not discovered by bears until the 26th, about 3 days after it was killed. On the morning of the 26th I saw the mother and 2-year-old cub 100 yards from the ram. The mother was moving about with nose to the ground. After about 5 minutes, her searching brought her downwind from the carcass and she walked directly to it. The mother bear fed for about 45 minutes. The cub spent most of the time playing with a piece of sheep hide. There was little left of the carcass except pieces of hide and bones. The mother did not bother to cover it with soil or debris as she probably would have done if more had remained. After a nursing, the family departed. This carcass was discovered by the bears too late for it to benefit them.

One day during the period when the bears were feeding on the old ram, a lone ram was seen on a low hill a little to the east. For 10 minutes he surveyed the country, as sheep are wont to do, before migrating across low and, from his standpoint, dangerous country. The ram then walked down the slope, followed a ridge to the stream bottom, climbed out on the tundra, and started moving across more gentle terrain. Along the way he startled small groups of caribou who were unaccustomed to seeing this white animal away from steep terrain and cliffs. A group of 30 caribou stampeded and circled back to gaze at the unperturbed ram. I watched the ram cross to safe cliffs. A wolf had been seen earlier in the day in the general area but apparently he had moved on. If a wolf had discovered this crossing, the ram might have become another bear cache. Crossings usually are safe because the low country is scrutinized carefully for some time before risking the crossing, but this illustrates the circumstances that might create carrion for bears.

The Bear in Possession Chases Away Smaller Bear

On 6 September 1959 an adult bear had covered a sheep carcass with sod and vegetation at the edge of a flat near Little Stony Creek. When I saw him, he was still pawing debris over the carcass, working slowly, as though disinclined to give up a pleasant activity. One or both of the forepaws were used to loosen the sod and scrape it and other debris toward the carcass. Sometimes the hindfeet also were used to push debris toward the carcass. When he stopped pawing, he lay on the heaped-up mound, the owner in possession. Later, he continued raking vegetation from all sides, forming a 40 foot circle around the carcass. Later still, the possessor galloped after a young bear that was about 150 yards away; chaser and chased both put on speed, and the young bear escaped easily. As the bigger bear walked back toward the carcass the intruder followed, sure of his ability to escape. Shortly before the owner returned to the cache, another chase took place and after escaping, the intruder lay down for a time before wandering away.

The following day very little was left of the carcass. The larger bear was gone and the small one was pawing about in the cache looking for scraps. He left with what remained of the head of the sheep, but dropped it on the slope so that I was able to retrieve it and learn that the dead sheep was a 10-year-old ewe. The cache area made a conspicuous dark spot on the flat.

Carcass at Little Stony Creek

Early in the afternoon of 7 September 1964, an adult female bear was seen stretched out a few feet from the carcass of a caribou bull. She lay on her side, occasionally raising her head to look around, and once stretching a foreleg into the air. After a time, she stood up to urinate and leave a scat. She surveyed the surroundings while standing on the cache, pawed a little more debris over it, walked away about one-quarter mile, and rested on a slope. An hour later she was roused by a young caribou bull, and stood up. Without hesitation, she walked rapidly toward the carcass. After pawing aside debris, she fed, pulling loose long stringy pieces. In a few minutes she pawed more loose debris over the cache and walked away again, passed the spot where she had been resting, and continued until she disappeared over a rise. The cache was about 15 feet long and 12 feet wide. I estimated about 25 bushels of sod and debris had been pawed over the carcass. The material had been scraped into an area about 30 feet across. The worn condition of the teeth indicated that the carcass was that of a very old bull (Fig. 42).

A Cache Unprotected

On 26 August 1956, in the Wonder Lake area just outside the park boundary, while searching for a caribou that had been shot and abandoned by a hunter, three companions and I came upon a fresh mound of vegetation consisting chiefly of sphagnum moss. The caribou carcass had been found by a bear. After a feast, the grizzly had covered the carcass with tundra moss. The cache in this case was not at all conspicuous, for the moss covering was green and brown like the surrounding

Fig. 42. This carcass of a caribou bull was mostly covered with sod and vegetation by a grizzly.

ground cover. A few days later, the carcass had been devoured. It is likely that this bear had not remained at the cache because of the human disturbances in the area.

Grizzly Feeding on Diseased Sheep

On the morning of 11 May 1964, some distance up a slope of Savage Canyon, a bear was pawing debris slowly over a sheep carcass. When I passed by 6 hours later, the bear was sitting on his haunches feeding on the carcass. On the following day the bear was seen lying beside the carcass, waiting for his hunger to return or at least for stomach space. A few days later at the site I found the sheep's hide, with four legs attached. Scattered about was much loose sod and debris which the bear had used to cover the carcass. The horns, detached from the skull but still attached to the hide, showed that the sheep was a 6-year-old ewe. The teeth were in bad condition. Two of the molariform teeth had been missing for some time, resulting in very long, little-worn teeth opposite the cavity of the missing teeth. The jawbone below the missing teeth was swollen and porous, the necrosis no doubt having caused the loss of the two teeth. The upper molars were worn down at an angle. Three incisors on one side had been missing for some time. One horn, worn smooth, had been broken off near the base, suggesting an old accident. A ewe with a bad mouth that no doubt affected her health had succumbed. The hide was bloody, suggesting that she might have been killed by a wolf or bear.

Prolonged Visitation to a Carcass

On 18 June 1961 a mother bear and yearling were feeding on the carcass of an old bull caribou that had become entangled in some wires left behind when the telephone line was dismantled. Two foxes were also present, waiting for an opportunity to feed. The following day I saw this family leave the carcass at 2 p.m., and cross a patch of overflow ice on the creek bottom before entering an extensive patch of dense willow. After their departure, Short-billed Gulls dived at a Golden Eagle perched on a ridge overlooking the carcass. At 3:15 p.m. a wolverine arrived and fed for 5 minutes. At 4 p.m. the mother and yearling returned and the mother fed for 35 minutes. She then pawed moss and other loose vegetation over part of the carcass, fed on a detached piece for 5 minutes, and lay down with her cub a few yards away.

For the next few days I saw no bears at the carcass, but a mother and two cubs were reported at the carcass on 24 June. In the morning of 25 June a large male had taken charge. He had covered the carcass with more sod and vegetation and rested on his side on top of the heap. Once as I watched, he raised his head lazily for a casual look, then relaxed, dropping his head to the mound. His sides moved up and down as he breathed (33–35 breaths per minute). Occasionally he adjusted his weight for more comfort. Toward the middle of the morning he fed drowsily on a jawbone. When I passed by at 7:30 p.m. he was still resting on the cache. On the 26th and 27th he continued to feed and sleep on the carcass. On the morning of the 28th I saw him climb a slope and disappear over the top. That evening a mother with two 2-year-olds was at the carcass, and on 3 July this family was seen leaving it. Only bones and a little hide remained. I thought the carcass was no longer an attraction, but on 18 July I happened to see a wolverine leaving it, dragging a leg bone, and on 4 August a mother and two spring cubs discovered the remains. She was pulling pieces of tendon from the long bones, tough tendons 15 or 18 inches long, which she swallowed without chewing.

Three bear families, a big male bear, two foxes, a wolverine, an eagle, gulls, and perhaps others were attracted to this carcass. Ordinarily, such an attraction lasts only a few days because it is finished in that length of time. I do not know why this one attracted attention for so long unless the wire entanglement caused some difficulty. After the big male grizzly left, I expect there was still an attractive odor, and small bits of dry flesh clinging to the entangled skull.

Lone Bear Dispossessed

One morning (30 June 1964) when I was studying vegetation on Sable Pass, I saw 50 or more caribou bulls, old and young, trotting past a young blond bear, about 3 years old. Two other bands were moving by him in such manner that for a few moments they were passing on three sides. Standing in the midst of these caribou movements the bear seemed uncertain whether he should become frightened. Finally, he made a short run, stopped, and then regained some composure.

In the afternoon when I returned to the pass, I saw this bear near where he had been in the morning, tugging at the half-eaten carcass of an adult caribou. Apparently it had been there in the morning, but it lay in such a way that I easily could have missed seeing it. The bear was pulling on the carcass as though trying to drag it up the slope. Later he managed to drag it about 25 yards down the rather steep slope. Three Golden Eagles were perched 50 yards away on the ground on the far side of the gully. One left, but the other two stayed for the half hour that I watched.

In the evening when I returned to the pass, the bear was pawing sod over the carcass. Two eagles were perched 100 years away. Soon the bear walked down to where the slope flattened out and drank at three small pools. In 5 minutes the bear started back toward his cache, galloping for a few yards, and was soon tugging halfheartedly at an exposed part. One eagle flew away. A mother with two spring cubs was on the west side of the pass. She had been in the area all day, and two lone bears also had been grazing on that side of the pass. They all were unaware of the banquet they were missing.

At 4:25 a.m. the following morning the mother bear with the two spring cubs had found the carcass and were resting; an eagle was perched 100 yards away; and the blond bear was grazing 150 yards down the slope. In 5 minutes the eagle took wing and I heard him calling as he flapped away. It would pay him to look for a ground squirrel rather than watch bears snoozing at a carcass. An hour later the family and the small blond bear were resting. Mew Gulls were in the area calling and again I heard the "wha-wha-wha" of the Golden Eagle. A few flakes of snow were in the air. By 6:40 a.m. a mixture of snow and rain was falling. The mother stirred, stood up, and pawed more sod on the cache. A little later the mother and cubs fed briefly and then all lay down. The snow was whitening the landscape.

On 2 July the carcass had been dragged down the slope another 50 yards where it had again been covered with sod. By evening the carcass was deserted—apparently all available flesh had been eaten. On 3 July the family and a lone bear were in the area but were not attracted to the carcass. On 4 July two gray wolves passed by while a small lone bear was at the cache. It chased one of the wolves that was carrying a leg bone as it disappeared over the skyline. Only odoriferous bones and hide remained.

Male Grizzly Discovers Moose Carcass

On the morning of 21 May 1963 a Mew Gull was diving at a Golden Eagle that was feeding on the carcass of an adult moose lying in the edge of one of the channels of the Teklanika River. In the afternoon I saw a chocolate-colored bear, a large male, at the carcass. He was biting the carcass and discarding mouthfuls of hair. A little later he removed and discarded some of the intestines. After feeding for 15 minutes, he pulled at the carcass, dragging it in the direction of the woods on the far side of the broad river bar. Clamping his jaws on the carcass, he braced all four legs and pulled backward, moving the carcass a few inches at a time. Once he rolled it over, and once turned it over end for end. He grasped the carcass at various places, sometimes pulling on a leg. At intervals he stood panting with mouth open. After dragging the carcass about 75 yards, halfway to the woods, he stopped his labors, stood briefly, and walked into the spruce woods, probably to rest.

The following morning the big male was lying on top of the moose carcass, which was still out on the river bar. No attempt had been made to cover it with gravel (the only material at hand). By early afternoon the carcass had been dragged just inside the spruce woods, and the male was lying on it. Out on the bar a small blond bear was nosing the spot where the carcass had been lying and the trail made when it had been dragged. The male saw the blond bear, and with long, slow strides came out on the bar a few yards, stood briefly watching the small bear walk away, then returned to his prize.

In the evening I found this male bear resting again on top of the carcass. A small dark bear moved from the bar into the woods a little distance north of the carcass. About 1 mile to the north the small blond bear was seen walking rapidly along the edge of the woods toward the carcass. As it came nearer, it raised its muzzle repeatedly to test the breeze. It walked into the woods and chased the small, dark bear, both disappearing among spruces. The male, hearing them, also disappeared into the spruces. The young bears reappeared on the bar some distance south of the carcass, the blond one galloping after the small, dark one, who continued northward for a mile.

The blond bear moved cautiously to the carcass, became nervous, and jumped away several yards. He circled into the woods and again approached the carcass, biting at it, then jumping away nervously. Before he could approach again, the big male returned and the blond disappeared in the woods. The male lay down on the carcass. I saw him on the carcass the next morning, but I could not find him a few days later.

Summary

Carrion in any form, fresh or ancient, is a special "delicacy" for a grizzly, although not a major food source. A large carcass may attract several bears over a short period of time, bringing them into closer contact than is usual. At most times, little overt strife results; larger bears have priority and others partake as temporary absence of a more dominant bear permits.

6
Grizzlies and Ungulates

Grizzly–Caribou Relationships

Large ungulates, when present, enrich the grizzly environment by contributing additional variety to the diet. Caribou dying from old age or disease are a sporadic source of carrion, as are the partially devoured caribou kills made by wolves. Although grizzlies are not slow, they are not fast enough to capture a healthy caribou that is beyond the age of early calfhood. However, a nominal number of very young calves do fall prey to bears (Fig. 43).

Each spring McKinley National Park caribou herds, numbering about 8,000 in 1962, after spending the winter along the north boundary of the park and northward toward Lake Minchumina, migrate through the park to Windy Creek country on the south side of the Alaska Range. After feeding for a few weeks in the Windy Creek area, they recross the Alaska Range, most of them traveling westward to the Thorofare River and beyond (Fig. 44).

Calving takes place between 15 May and 15 June, chiefly during the period that the caribou are moving toward the south side of the range. The time of migration and the route may vary from one year to another, so that the height of the calving season may occur in different localities in different years. In 1939, for instance, most of the calves were born near Wonder Lake. By the time the herds had reached Polychrome Pass and Teklanika River, most calves were old enough to be no longer vulnerable to a grizzly attack. The next 2 years, calving took place far to the east, much of it between the Teklanika and Savage rivers, giving the bears in that area an opportunity to capture young calves. In 1965 many young calves were present between Sable Pass and Toklat River. Hence, the bear inhabitants in this section of the park were thoroughly calf-conscious for a period in late May and early June.

Fig. 43. Robust caribou such as these almost never supplement the grizzly diet.

Fig. 44. Migrating caribou.

I should add that I noted no special movement of bears into a calving area for the purpose of preying on calves. If the calving took place in one area consistently and was concentrated more than it is, one could conjecture that some bears develop a movement pattern that takes them to a calving ground for the annual calving period. Bears ranging during any calving season where caribou are scarce or absent probably are not aware of missing anything, and subsist on other springtime fare.

Grizzlies in calving country are aware of the potential vulnerability of calves, and may be seen chasing bands of caribou on the chance that a calf too young to escape will falter and fall behind. Caribou bands generally are chased indiscriminately, perhaps often without the bear knowing whether a calf is present. Bears hunt on a percentage basis. If the season is right, fortune strikes sooner or later and a young calf is overtaken.

Newborn calves gain rapidly in strength and within a few days of birth are strong and fleet enough to escape the grizzly. As the calving season wanes, hunting success drops. The meat-hungry bears may continue to chase caribou for a time after the season for weak calves has passed, but a series of failures soon discourages them. Once they recognize that chasing caribou is no longer profitable, they resign themselves to the inevitable, forget about calves, and for the rest of the summer pay little heed to caribou.

Below I describe some of the behavior and hunting incidents involving grizzlies and caribou, chiefly during the calving period.

Alert to Calf Possibilities

On 6 June 1961 a mother and her blond yearling cub moved down a slope of Igloo Mountain. She must have been hungry for meat because she was especially interested in caribou encountered as she moved down the slope. When she saw three caribou bulls feeding in tall willow brush some distance below her, she stopped and watched for 4 minutes. After moving down a little farther, she again stopped to watch the bulls, now only about 60 yards away. The bulls discovered the bears and trotted away. A little later she caught the scent of caribou, stopped, raised her muzzle to test the air, then proceeded at a fast walk. The yearling cub, seemingly aware that his mother was bent on a hunting project, remained behind resting on his stomach with nose between paws, watching the mother advance. After traveling forward a hundred yards, the mother stopped, raised her muzzle to scent the breeze, then stood erect on hind legs to look around. Soon a large bull caribou 25 yards away became aware of her and trotted away briskly. The bull did not interest her. The cub seemed to decide that the incident was finished and galloped down the slope to join his mother. Both moved forward, the cub in the lead. The mother then reached forward and pushed the cub aside with a paw and loped into a hollow grown over in tall willow brush where she and the cub became hidden from my view. Another bull caribou emerged from the hollow, stood looking into the hollow, and seemed uncertain of the position of the bears. In a minute he noted the proximity and galloped away. The bears were able to make a close approach because the breeze favored them. Lower on the slope, two more bulls were similarly flushed out of a draw. The mother bear seemed especially hopeful that an opportunity for capturing a calf would develop but there were no cows or calves on the slope. She then turned to feeding on roots and grass.

Apprehensive Caribou

The behavior of caribou varies on different occasions. One day caribou may be complacent when near bears, and on another occasion they may be especially timid. On 3 June 1955 I discovered a band of 30 caribou hurrying away from a bear standing 300 yards to their rear. Possibly the bear had been chasing them and they were continuing their flight, taking no chances. A band of 100 caribou was hurrying away from another bear who was walking toward them far to their rear. Perhaps they too had been chased. Later on the same day I saw a small band galloping away from a bear, but the bear was making no threat to pursue. All these groups consisted chiefly of cows and calves so they had cause to be prudent.

On 20 May 1961 three migrating caribou cows showed more than usual concern on seeing a mother bear and yearling. The cows stopped 300 yards from the bears, watched briefly, then changed their course so as to pass far to one side of the bears. None of the cows had a calf so their extreme wariness seemed unnecessary. Their concern may have been associated with the season—calving time for one or more of the cows may have been imminent, causing behavior appropriate to the presence of a vulnerable calf in bear country.

Only Slight Reaction to Grizzlies

On 26 May 1961 a blond bear on an old river bar took a course parallel to and about 150 yards from a herd of 250 caribou, among which were many calves. The caribou seemed to take little note of the bear; only two or three cows seemed at all concerned. The long line of caribou drifted slowly, almost imperceptibly, a little to one side of the bear's line of travel. I expected the bear to make a try for a calf but instead he moved a short distance beyond the herd and, during the hour that I watched, dug roots.

The following day, on the same river bar, I saw what appeared to be the same bear approaching another large herd of caribou. A short distance from the herd he stopped and fed for 25 minutes in one spot, probably on the remains of a calf. He then walked parallel to the herd and not far from it. Several caribou that were lying down stood up to watch the bear, but most of the herd paid little attention to him and continued grazing. I noted several calves near the far end of the herd and thought the bear might make a try for one, but instead he chased and captured a ground squirrel, then fed briefly on roots before moving slowly up and over a low hill. The bear was apparently not very hungry for meat. Of course, he did capture a ground squirrel, but no bear can resist chasing a fleeing squirrel. Feeding on roots on these two occasions suggests that bears prefer a mixed diet.

On the morning of 6 July 1948 over 200 caribou were resting in a sedge meadow near the Toklat River. Beyond the caribou, near the base of a slope, I discovered a mated pair of bears also resting. During the course of the next few hours, the bears crossed the meadow three times at the spot where the caribou rested and fed. On each occasion the caribou merely moved aside enough to form a lane for the passage of the bears. A few times the female veered slightly toward the caribou, just enough to cause a slight widening of the corridor. The caribou seemed especially serene, perhaps due in part to the fact that the calf-hunting season had been over for several weeks. Furthermore, the bears were concentrating on mating (the herd of caribou probably had been accustomed to the maneuvering of the male and female).

On 4 July 1963, in the same area, a herd of about 300 caribou moved slowly to a snowfield up the slope when a bear passed. The caribou needed little stimulus to make this movement because the flies were beginning to bother them; they were already on the verge of retreating to the snow.

It often appears that caribou in a herd are less wary than when alone or in a small group. In a large herd responsibilities do not rest to so great an extent on each individual. (In a string of pack horses it is the horse in the lead that is alert and watchful. Put him farther back in the string and he becomes a follower.)

Futile Chases During Calving Season

During the calving season, there are many futile chases either because there are no calves in the fleeing herd or because the calves are old enough to escape easily.

On 12 June 1948 an optimistic bear spent 50 minutes chasing a group of caribou without results (Murie 1951).

On 1 June 1955 I saw a band of 30 cows and calves trotting easily along Polychrome Flats. Far in the rear, so far behind that it appeared to be a separate incident, was a mother bear galloping after the departing caribou. Over to one side were six caribou standing gazing at the passing bear. The grizzly finally quit the hopeless chase and backtracked toward two yearling cubs left three-quarters of a mile behind her on a gravel bar where, I assume, the chase had started.

On 28 June 1956 a grizzly galloped his fastest in the wake of about 20 caribou that quickly left him far behind.

On 23 July 1963 a bear was seen traveling quite near about 250 caribou. When the bear ambled a little toward the herd, there was a slow movement away by those animals nearest the bear. He could not resist chasing the herd which hurried up the nearby slope. Then the herd parted and came down the slope on either side of the bear. The bear reversed his direction but soon realized the chase was hopeless.

In the evening of 4 July 1949 four of us climbed a steep slope along the Toklat River to classify a large band of ewes and lambs. On an open grassy slope directly below, we saw about 200 caribou. A few near the edge of the woods attracted our attention by their running, and behind them we saw a mother grizzly and a 2-year-old cub emerge from the woods. The mother led the way, walking stolidly into the open. The caribou fled to either side as she progressed. When they came to the middle of the meadow, the cub could restrain himself no longer and galloped pell-mell down the slope after the caribou. In a twinkle they had disappeared into the wood and the cub was alone and out of sight of everything, including his mother. He stood on hind legs and walked seven or eight steps trying to see her; she sat erect on her haunches looking for the cub, but apparently less concerned than he. He probably was a little disconcerted to find himself alone. He started back the way he had come, was soon in view of his mother, and travel with her across the grassy slope was resumed.

A Very Young Caribou Calf Captured

On the morning of 24 May 1961 I watched three cows and a calf caribou trotting westward on Polychrome Flats. They were headed in the opposite direction from the eastward, spring migration. Looking for a possible explanation, far to their rear I saw a lone bear following at a steady gait and putting his nose to the ground repeatedly as though following a trail. I am not sure that he had seen the four caribou.

A little later a herd of 13 cows and 4 calves passed a little to one side of him, and he loped after them. Soon five of the cows without calves swerved sharply and the bear cut across after them, apparently unaware that no calves were present. In his experience a calf could be present in any group and he followed his customary routine of chasing the nearest group without trying to determine if a calf were present. The five cows made another sharp turn and stood watching the bear approach. He disregarded them and continued forward toward a cow and calf not discovered previously, standing a few hundred yards away. On seeing the oncoming bear, the cow, followed by the calf, trotted away slowly.

Evidently the calf was very young, for after traveling a short distance it lay down and the bear soon reached it. The mother ran in an arc and joined the five cows that had been watching the bear pass by, and the group moved eastward again. I expect the mother later returned to the scene. The bear fed on the calf for 40 minutes, then walked 100 yards away and disappeared in the rain and fog that had moved in.

Within 2 miles of this bear there were two other bears playing (probably 3- or 4-year-olds) as they traveled toward a slope; two more young bears about the same age were digging roots; one lone bear was traveling and another feeding on a carcass. However, only the one incident was observed during the morning, although many caribou were moving through the area where these seven bears were active.

Sympathetic Cow

Often one finds a worried mother caribou attending a bear that is feeding on her calf. She may remain nearby for a day or longer. On 2 June 1962 a mated pair of bears was feeding on a calf on tundra, near the East Fork River. Two caribou cows were circling and watching the bears helplessly. The second cow probably was a "sympathetic" companion who had joined the distressed mother. I have seen this behavior in other instances among caribou and once I saw a cow moose join another cow moose whose calf was in distress.

Food Preferred to "Love"

On the morning of 31 May 1963 I saw a large, dark male following at least half a mile behind a female, so far behind that she was out of sight, forcing him to follow her trail. He was patient and moved along slowly. She was on the prowl, no doubt hungry. After traveling eastward some distance, she turned south toward a band of about 90 caribou scattered and feeding on foothills one-half mile away. When she neared the caribou, she broke into a gallop. The caribou fled, hurrying off in a compact group. Quite soon the female bear stopped and apparently captured a calf that was too young to escape. She fed on the calf until the male, still moving deliberately, was about 200 yards away. She watched him approach then picked up the carcass and galloped up the slope with it dangling from her jaws, and disappeared behind hills. In time the plodding male, following her trail, also went out of view. The female obviously was not ready for mating.

Family Discovers Calf

On 8 June 1964 I saw a mother and yearling that I had watched frequently the year before, moving in an irregular course. She made a half-hearted, short dash at a lone caribou and then continued her irregular course, the yearling sometimes trailing 100 yards behind. I watched them for 1½ hours hoping to get pictures of the cub showing its size. They settled down to digging roots on a slope for 15 minutes, then moved out of view over the top. A few moments later a caribou cow emerged from the spot where the bears had disappeared. It was using a fast, swinging trot, and obviously was startled by the bears. I climbed to where I could see the other side of the slope and saw the two bears feeding on a calf. There had been no chase, for the bears were feeding near the point where they had disappeared. The calf either was too young to escape or had been stillborn. The cub fed for half an hour and then rested against the side of the mother who kept pulling off little pieces and chewing them thoroughly. She fed for an hour before walking over to a snowpatch for a few bites of snow. The bears lay down out of sight of the calf remains. An hour and a half later, when the mother caribou moved up close to the carcass, the mother bear took a few steps to where it could see the mother caribou, looked briefly, yawned three times, walked back over the rise, and lay down again.

Three days later, while hiking up on Primrose Ridge, three of us came upon a caribou cow that also ran off with long, swinging strides, so effortless she seemed to be floating. When we passed that way later she paced away from the area as before. We searched and found her dead, newborn calf, perhaps stillborn. Our experience was similar to the bears'. If our sense of smell had been keener we would have found the calf when we first saw the cow.

A Bit of Inference

On the morning of 9 June 1964 I watched a bear in the distance walking steadily eastward until he came downwind from a group of seven resting caribou. Catching their scent, he made a right angle turn and lengthened and quickened his stride. He continued walking toward the caribou for 200 yards, and when he was about 75 yards away, they stood up and swung away to the west. The bear galloped toward the beds they had left, perhaps hoping that a calf had been left behind. He gave the spot a rather perfunctory sniffing and resumed his original eastward course. The caribou stopped, turned, and trotted after the bear, and when to one side of him stopped to watch him.

Old Male Attracted by Tumultuous Stream Crossing

One afternoon in early July 1947, several hundred caribou were resting on the gravel bars of Thorofare River. I approached and took movies as they splashed across the main channel of the river. Many of the calves lost their footing in the swift, glacial stream and had to swim. There was prolonged turmoil as scores of caribou crossed the stream and climbed a steep bank to the extensive green benches beyond. Then I noticed a big, male grizzly that was grazing on the opposite side of the river and had become aware of the commotion and excitement. Apparently, he concluded that this was an opportunity for food. He may have thought that the caribou were being attacked and one might be killed rather than expecting to catch a calf himself, for the calf-hunting season had been over for several weeks. He galloped ponderously down the slope off the green bench on which he had been feeding and continued across the gravel bar. I retreated, but he seemed unaware of my presence. By the time he reached the stream the last of the caribou were crossing. He plunged into the rapid current and pawed his way up the steep bank that had been made wet and slippery by the dripping herd. The caribou were all out of reach as their swinging gait carried them gracefully over the rolling tundra. The bear surveyed the situation, then grazed in a lush green hollow.

In May and June 1965, the drifted snow and late spring delayed the eastward migration of caribou so that between Sable Pass and the Toklat River caribou were present in larger numbers than usual during the height of the calving season. The grizzlies in the area were well aware of the calves so for a period the opportunity to witness grizzly–caribou behavior, as well as some wolf activity, was plentiful, and resulted in the following observations.

The Caribou Calf Season in 1965

I spent the first day of June 1965 in the Polychrome Pass area. A bear was first seen at the head of one of the branches of East Fork River. For about 10 minutes he fed in one spot, apparently on part of a carcass, then wandered southward. Twice he stopped to roll. There were no caribou ahead of him, so I moved on (Fig. 45).

Farther along I saw a lone bear traveling down to the Polychrome Flats, where small groups of caribou were feeding or traveling, and pass to one side of five caribou. They were unaware of each other because of the direction of the wind. He continued forward and when he was within 200 yards of a herd of 14 cows and 1 calf, he loped easily toward them. The caribou galloped away and dropped off the bench to a gravel bar, where eight of the cows swung off to one side and stopped to watch the bear as he galloped after the other six cows with one calf. The calf must have been quite young because it soon fell behind the fleeing cows. When the bear saw the calf fall behind, he put on extra speed and soon captured it. It had circled sharply and given the bear an additional advantage. He carried the calf 50 yards across the gravel bar, onto the dwarf birch-covered bench and started feasting. I moved on to look for two wolves, and when I returned an hour later, the bear was lying near the remains of the calf waiting for digestion to make room for more. He raised his head at intervals to look around—his only concern was another bear.

Fig. 45. The flats south of Polychrome Pass where bears seek caribou calves in some years.

The mother of the captured calf returned and was maneuvering anxiously 100 or 200 yards from the bear when a second cow joined her and they moved about together. One-half mile away a lone cow with a calf approached and passed about 100 yards from the bear, unaware of her precarious location. The bear was sleeping off its big meal, and neither it nor a wolf resting nearby was much interested in calf scents.

About noon I discovered a lone bear to the west, chasing a group of five cows and three calves. It was a long chase but the calves were still holding their own fairly well after about half a mile. The chase passed several hundred yards south of the resting bear, then turned and came down the gravel bar toward the spot where the resting bear had made a capture earlier in the day. It had been a long run for the calves, and they were scarcely holding their own, about 100 yards ahead of the pursuing bear. The resting bear either heard or scented the chase to the east, for he stood up, galloped to the edge of the brushy flat, dropped down 10 or 15 feet to the gravel bar, and angled out to intercept the pursuing bear. Both were partly hidden by the bank when they met, but apparently there was a slight altercation. The one pursuing the caribou was the aggressor and the "local" bear galloped farther out on the river bar to escape. Before being overtaken, however, he stopped, and the two stood "glowering" about 15 feet apart. After a brief about face, the pursuing bear retreated to the west at a gallop and found the remains of the dead calf belonging to the other bear. It grasped the carcass in its jaws, started galloping, dropped the meat, and continued galloping toward the spot a mile away where I first saw it chasing caribou.

Far ahead I saw another bear watching the oncoming one. Its size could not be determined immediately, but as the galloping bear drew nearer it obviously was the well-known, dark mother hurrying back to her 2-year-old cub who had been left behind when its mother had chased caribou. As the mother neared him, the cub seemed uncertain of her identity and galloped a short distance away before coming forward to meet her. The mother, although hungry, was so concerned over her cub that she had dropped the carcass of the calf and hurried back to it. The altercation with the other bear on the river bar may have heightened her solicitude.

After joining her cub, the mother walked back toward the calf carcass, and the cub followed closely. When they reached the carrion, the two bears fed for 2 hours and then a nursing took place, after which the mother resumed feeding on the carcass and the cub rested on its side.

I stopped watching at 2:50 p.m. When I returned at 4 p.m., a herd of about 500 caribou was feeding on an old river bar far up toward the head of the west branch of the East Fork River. About a quarter of a mile beyond the caribou, the mother bear and her 2-year-old cub were traveling parallel to this spread-out herd, the cub romping ahead of his mother. They had traveled over 2 miles from where I had left them a little over an hour earlier. Occasionally, they stopped to dig a few roots. In the next 2 hours they moved around the herd in a long arc. A group of 20 caribou, including one calf, that the bears encountered moved off to one side. The cow with the calf was most circumspect for she moved out in front of the others. The bears went part way up a slope and lay down. After 10 minutes, the mother walked 20 yards, rolled over on her back, and the cub nursed. It was 6 p.m., and at 7 p.m. when I left the bears were still resting.

While these observations were under way, other dramatic activities were in progress. At 4:30 p.m. another lone bear, with nose close to the ground, moved about in circles over a sedge meadow. He moved into some gently undulating ground where I caught occasional glimpses of him as he continued working to unravel a trail. A lone caribou was flushed out and the bear galloped toward her but stopped where she was first seen, circled as before, but soon moved out of view. Fifteen minutes later he was working the sedge meadow again, but was soon back where the cow had been flushed out, and this time encountered a cow with a very young calf. A chase of 300 yards ensued, and the calf was captured easily. The bear had trailed persistently for 1 hour and 15 minutes, covering an area less than one-half mile in diameter. This persistent trailing to find a calf was seen rarely.

In the stretch of country between East Fork River and Toklat River, about 900 caribou, one bear family, and four lone bears were seen during the day.

On 2 June 1965 at 6:10 a.m. in the Polychrome Pass area where I had watched caribou and bears the previous day, I discovered a bear carrying a calf up a steep snow bank to the bench above the river bar. Apparently, the calf had just been captured because 14 caribou were trotting away from the scene, and a lone cow was maneuvering anxiously near the bear.

About 2 miles from this bear, another lone one was seen first striding along and then stopping to feed on a carcass. He may have known about this carrion and was returning to it.

Later, at 8:45 a.m., the mother and 2-year-old cub seen the previous day were discovered chasing rather half-heartedly a small band of caribou, and the mother soon gave up. As they traveled, they encountered four or five small bands of caribou but did not chase any. After about an hour, they veered to one side and fed for 10 minutes on a carcass they had discovered, then continued on their way.

At 6 a.m. on 3 June 1965 my attention was drawn to 15 caribou below us, galloping eastward on Polychrome Flats. They obviously were fleeing from a wolf or bear. Scanning the country behind them I discovered a grizzly 300 yards away, galloping in pursuit. Presently, he stopped and for about 10 minutes meandered about with nose to the ground over a patch of tundra 200 yards in diameter. In one spot he seemed to find a morsel that detained him for a couple of minutes. Finally, he dropped this project and walked out over the tundra, occasionally nipping at dwarf birch and willow tips.

When he came to a snowfield on a slope, he followed a bear trail to the top of the drift, then reversed his direction and followed the trail to the bottom, stepping in the tracks of the trail all the way up and back, not missing a single one. A lone cow caribou, lying in the tundra a little over 100 yards downwind from the bear, was watchful but the bear was unaware of her. She probably had a young calf nearby.

After lying in the snowfield for a short time, the bear walked steadily across the tundra for three-quarters of a mile. He galloped toward six caribou 200 yards ahead of him. They sped away and he turned aside toward four others who circled and then watched him from 75 yards away. There were no calves in either band. The bear wandered off and when I left at about 8 a.m., he was digging roots. Scattered, traveling caribou moved to either side of the bear as they passed by.

I saw three other lone bears during the morning. I watched one feed on crowberries on an old river bar. Another, that was traveling across a gravel bar near the head of a river, startled seven or eight caribou that trotted in a tight circle and stopped to watch him go by. A third bear was seen in foothills, feeding on crowberries.

In mid-afternoon I discovered the dark mother and her 2-year-old cub on a snowfield east of the East Fork River. They were moving east along a draw when they became aware of a herd of 30 or more caribou on the slopes about 300 yards above them. The mother started loping up the slope toward the caribou, the cub following 30 or 40 yards behind. The caribou hurried and were quickly out of view and far away. The bears continued loping up the rather steep slope for perhaps a quarter of a mile before the mother stopped. The cub caught up to her, and, enjoying the chase, loped ahead. The mother followed for a short distance then stopped, and both bears climbed out of view into a deep draw on the other side of a ridge. Soon I saw another group of caribou climbing a slope and surmised they had been startled by the bears. In their hunting the mother and 2-year-old had moved 3 miles beyond their usual range but the following day were back in their old haunts.

During the day, I had seen four lone bears on Polychrome Flats, and two families and two lone bears on Sable Pass. About 600 caribou had been seen in the area.

On 4 June about 1,300 caribou, in groups of various sizes, were distributed widely in the Polychrome Pass area. Four wolves spent the day resting on a gravel bar. During the day, five lone bears and a mother and cub were observed but their activity was so scattered that I missed some.

The dark mother and her 2-year-old cub, seen over the previous few days, were again on hand to exploit calf-hunting opportunities. I saw the family at 8:30 a.m. traveling on a gravel bar. As the mother became aware of six caribou cows, she and the cub loped toward them. There were no calves and the bears soon abandoned the chase. They walked across a gravel bar, climbed a bank, and, a short distance out on the dwarf birch flats, stopped to feed on a calf carcass. A concerned cow circled nearby, apparently the mother of the dead calf. The bears fed for 15 minutes, moved off to feed on crowberries for 5 minutes, and returned to the calf carcass where the mother bear rested and the cub chewed on remnants for one-half hour, after which the cub nursed. They both fed again at the carcass, apparently cleaning up the remains, before departing to rest on a snow bank for 1½ hours. At 11:25 a.m. these bears became aware of a lone bear sniffing about where the calf carcass had been. They galloped to the top of a gentle rise to watch the lone bear who, finding nothing to eat, continued on his way. The mother apparently recognized the bear as one too small to worry about because she moved back to the snowfield before it left and nursed the cub again. They rested until 12:30 p.m. and then walked south a mile, watched a band of 15 caribou pass by, and fed on crowberries. In an hour they were back sniffing the spot where the calf carcass had been and then continued traveling.

On a flat of dwarf birch they came upon a herd of 40 caribou including 5 or 6 calves resting. The caribou sped away, but soon stopped to watch the bears loping toward them. Two of the cows with calves watched the mother bear until she was quite near. The mother bear must have been aware of the presence of calves for she increased her speed. One of the fleeing calves, for whom the dwarf birch brush made motion especially difficult, could not keep up with the others. When it reached a grassy lane, it managed to stay ahead of the bear but was captured when it turned aside. The 2-year-old cub was left far behind and had some difficulty finding its mother. When he located her, he approached cautiously and stopped often to stand erect on hind legs to look. The mother never once looked up to assure the cub, but fed hungrily. They both fed for three-quarters of an hour, then walked to a snowfield where the mother quenched her thirst with seven or eight mouthsful of snow. She loped easily toward a cow and calf that were passing but only followed a short distance. After a session of play, the bears returned to the carcass to feed and rest, then moved to a small stream where the mother drank and then returned to the carcass. When I left at 5:30 p.m., the mother was resting and the cub feeding. Later in the evening, I did not see the bears; apparently they were resting out of my view. The mother had captured one calf during the day, and possibly two.

At 7 a.m. the same day we saw a lone bear digging roots on a bar toward the head of East Fork River. A little later this bear started traveling and by 11 a.m. had moved 5 or 6 miles in a large loop and chased—all without success—10 groups of caribou ranging in number from 3 or 4 to 150. During this time he did not feed.

Another lone bear seen early in the morning chased 200 caribou that doubled back around him and settled down on a river bar to feed and rest. Later he was seen chasing five bands of caribou, one after the other, without any luck. All these chases were similar—the caribou sped away rapidly, leaving the bear hopelessly behind.

Three other lone bears were seen briefly as they rested, traveled, or fed on crowberries and roots.

It was apparent that the bears were having more difficulty capturing the calves who were now becoming old enough to escape easily.

About noon on 5 June 1965 a lone bear was observed walking steadily toward a herd of about 150 caribou. When 200 yards from them, his walk changed to a slow lope. The caribou were not alarmed, for as he came close they moved off a short distance to either side and stood watching, thus forming an aisle for his progress. After loping easily through the middle of the herd, he spotted two cows each with a calf at the edge of the herd and shifted to high speed. One of the calves soon fell behind the other three caribou and was soon overtaken. After feeding on the carcass for an hour, the bear started walking in the direction from which he had come, toward foothills 2 miles away. En route he encountered a group of 20 caribou, loped after them until they turned to one side, and then resumed walking as before. Reaching the foothills, he climbed far up a slope and lay down on a rocky outcrop. This long retreat from the carcass to the cliffs suggested unusual caution. This behavior was like that of bears with cubs in spring.

On 6 June at 2:45 p.m. I discovered a blond mother and her blond yearling near the head of the west branch of East Fork River. They were traveling west and climbed one of a series of parallel ridges that come off the Alaska Range and terminate on the south side of Polychrome Pass. In the next 3 hours they climbed and descended four of the parallel ridges and climbed the fifth, traveling steadily except for one stop of 15 minutes to dig roots. Along the way they encountered several caribou. Forty caribou resting to one side of their path were startled and ran down the slope to one side, but the mother bear paid no attention to them. Coming over one ridge she frightened a herd of 30 caribou, made 3 or 4 jumps toward them, and resumed her travel. Near the top of the fourth ridge she encountered six cows and a calf and chased them to the bottom of the slope. Near the top of the fifth ridge, about 50 caribou were feeding in a basin. As the bear swung to the left, toward the caribou, I shifted from field glasses to telescope. During the few moments this took, she dispersed the caribou, and through the telescope I watched her walk up the far slope of the basin and stop to feed on a carcass. She was so far away that details could not be seen. A worried cow came over the top of the ridge and stood watching the bears feeding, apparently on her calf which had been too young to try to escape or had died earlier. The bears fed for about 25 minutes and then rested nearby.

The following morning, 7 June, I saw the mother and blond yearling a mile from where they had been feeding on a calf the previous evening. The mother of the calf was still near where the carcass had been. Soon after I saw the bears, they turned and loped 200 yards to feed for 10 minutes on a carcass, probably the remains of a calf. The family then traveled across the dwarf birch flats, startling a lone cow who pranced ahead of them with tail erect and twice sky-hopping. The bears ignored her. At 7:30 p.m. the bears galloped at right angles to their course toward two cows who circled and trotted up close to the bears, then trotted off with tails erect. The bears sniffed around where the cows had been but found nothing. Farther on, five more caribou watched the bears who seemed to have found the remains of a carcass for they fed for 35 minutes. One hundred and seventy-five caribou were scattered about on the flats.

On the afternoon of 7 June 1965 a lone bear loped after 15 cows and 6 calves that quickly left the bear far behind. Soon afterward this bear encountered a lone caribou and chased her briefly. Later in the day we saw a lone bear chase three cows for a short distance. A few chases occurred during the day but there were no captures. During the following days that the bears were observed, a few chases occurred, but no kills. Most of the calves were apparently strong enough to escape the bears and, as the 1965 calf-hunting season was over, they had to wait until next year before again hunting calves (Fig. 46).

Every spring the drama of grizzly bears supplementing their vegetarian diet with young caribou occurs in the tundra. This is an old relationship, one of the natural ecological activities still existing in McKinley National Park. Those grizzlies living on the migration route that caribou use during the calving season capture a nominal number of young animals, but even in years when grizzlies find exceptionally good caribou hunting, their activities have little impact on the caribou population.

Grizzly–Moose Relationships

A mother moose with one or two calves is formidable, even to a grizzly. During the last half of May and early June (the calving period) the grizzly, and also the black bear, consider a moose calf potential food. However, the mother moose is not easily daunted and can often discourage a bear by her belligerent presence. Individual behavior varies in both moose and bears when confrontations occur. I expect that a large male grizzly is less easily deterred than is a smaller bear (Fig. 47).

Fig. 46. Caribou seeking snow to minimize the attack of botflies and nose flies. The large herds furnish carrion for bears.

Fig. 47. Mother moose followed by a very young calf in bear country. Bears occasionally capture young calves.

The mother moose may slip away prudently with her calf when it is a few days old to avoid a bear she has discovered in the neighborhood. There is, of course, the possibility of a bear finding an unguarded calf and capturing it before being discovered by the mother. In Wyoming in mid-June Conley (1956) saw "a black bear carrying a squealing calf moose in his mouth. Almost immediately a cow moose appeared and attacked the bear. She jumped on the bear's back, striking him with her front hooves. The bear dropped the calf and turned to fight the moose." Conley shot the bear, ending the incident. A little later the cow and calf could not be found. The bear was 6 feet 9 inches in length and weighed 350 pounds. The moose had inflicted a deep gash in the bear's shoulder with her hooves.

The belligerence of the cow moose is illustrated by an interesting observation made by Altmann (1956) about a band of horses seen in June swimming to an island in the Snake River (Wyoming) to feed.

> A few minutes later two more heads were showing in the water, but it was apparent that they did not proceed without difficulties. In fact, they seemed to collide, and the field glasses revealed that one of them was a horse, the other one was a moose cow trying to hinder the horses from landing on the green island. A serious battle ensued with the horse being ducked and rapidly losing in speed and strength. Eventually (after about 12 minutes) the horse managed to climb ashore, staggering and tried to graze. The moose, ears folded back, turned to swim to the other shore in swift strokes, and disappeared in the willow thicket. It can be assumed that the horse, in passing to the river banks, came too close to a moose calf and that the moose cow became aroused.

The respect that grizzlies and moose have for each other tends to keep them apart after the calving period. I have no record of a grizzly killing an adult moose in the park, but in Wyoming others have reported bears killing adult moose in the spring, floundering in deep, packed snow through which their long legs break and slide downward at unpredictable angles that cause them difficulty. A short, vigorous struggle of this kind might soon exhaust a moose weakened by a long, hard winter, but ordinarily such snow conditions are not known in McKinley National Park.

The effect of bear predation on moose populations is difficult to determine, but moose in McKinley National Park have prospered in spite of a large grizzly population. Predation on moose calves is sporadic because calving takes place over a vast acreage, and even a bear trying to kill a calf seems infrequent. Some bears are busy with their root digging and seem unaware of the potential.

The following incidents that I have observed in McKinley National Park illustrate the behavior that may occur when grizzly meets moose.

Cow Moose Keeps Bear at Bay

Elsewhere (Murie 1961) I have described the behavior of a moose guarding her two calves, and how she kept a large grizzly at bay successfully.

Bear Captures Calf

In June 1962 a bus driver saw a bear chasing a cow and calf along Igloo Creek and watched the bear capture the calf and drag it into the brush. The cow continued running, not realizing that her calf had been captured. No other details were observed because the bus continued on its way. This incident was complicated by the appearance of the bus on the scene and the retreat of the cow.

Large Bear Kills a Calf

On about 1 June 1961 a park workman saw a bear that he thought was a big male kill a calf. His car almost ran into a cow moose with two calves and a bear chasing one of them. He backed away from the animals and the conflict moved toward him as he continued to back away for "about a mile." The cow struck at the bear with front feet, and the bear stood on hind legs to strike the moose with a paw, but apparently no contact was made. The bear finally succeeded in killing one calf and carried it into the brush near the road.

Cow Moose Retreats From a Large Male Grizzly

On 21 May 1961 I saw a cow moose with two newborn calves and, 60 yards away, her unwanted yearling. They were in a willow depression near the base of a steep incline on Polychrome Pass. It seemed an ideal nook, away from travel routes, in which to be sequestered with the calves until they gained strength.

On 26 May a lone, 3-year-old grizzly moved down the steep talus slope above the moose. The noise of rolling rocks alerted the mother. When the bear saw the moose, it turned at right angles and, with what seemed a cautious and watchful attitude, followed a contour above. When he had passed the moose, he broke into a gallop and descended the slope at some distance to one side of the moose. His anxiety caused him to continue galloping for half a mile after reaching the flats at the base of the slope. When the yearling moose saw the bear, it took fright and hurried down the slope to the flats, moving out on the tundra. The cow and her twins moved only about 30 yards down the slope. Apparently this mother had estimated correctly the size of the bear for she did not seem frightened or even apprehensive.

Three days later, on 29 May, the mother moose and her twin calves were in the same retreat. Out on the tundra a large male grizzly was on the move, no doubt in search of a receptive female. He disappeared from view on the river bar below me. When I saw him again, he had climbed the slope and was on the edge of the little basin where the moose family was staying. As I watched the bear moving forward, only 40 yards from where the family had been resting, it appeared that a serious altercation would take place. But about that time both the bear and I noticed the cow and two tiny calves climbing the long steep talus slope leading up to the road. The cow apparently had moved off as soon as the bear came to the edge of the basin, or perhaps a little before. Because of the willow growth and the lay of the land, it is possible that she had seen the bear when she started to leave, but she also may have scented him and started to leave before he was in sight. The yearling moose was not present. The talus and large rocks and the steepness of the slope made climbing difficult for the cow and calves. The bear started climbing after them but they had a good start and were able to reach the road while he was climbing slowly, far below. The family moved down the road away from my car; when the bear reached the road, he came toward me instead of following the moose, and on seeing me climbed to the skyline above and disappeared.

The reaction of the moose to the big male was far different from her behavior at seeing the small bear a few days earlier. In the second incident she seemed to know that a large bear was approaching and retreated discreetly with her calves.

Male Grizzly Feeds on a Young Calf

On 25 May 1962 I watched a 3- or 4-year-old bear near the top of a pass between Igloo Creek and Big Creek. He crossed several low ridges and followed one leading toward a cow moose, who stood immobile except for cocking her ears occasionally toward something on the ground nearby, no doubt one or two very young calves. The bear stopped abruptly about 75 yards from the cow, stood watching her briefly, then retraced his steps, recrossing the ridges over which he had come. A little later I saw a calf beside the cow. A yearling moose appeared from the direction the bear had gone and trotted to a point 150 yards from the cow moose who, no doubt, was its mother. The cow climbed toward the yearling with ears held down threateningly and angrily, and when she was near, she ran and struck at it with front hooves as it retreated. Later, after she had returned to the calf, she again chased the yearling who did not wish to leave its mother.

About one week later (2 June) I saw a large, male grizzly lying where I had seen the cow moose with her calf. The cow was moving about anxiously in the vicinity, coming to within 40 or 50 yards of the bear, who would raise his head slightly each time she approached. This situation continued from 5 p.m. to 10 p.m. Once the bear walked down to the creek for a drink but returned to his resting spot. The cow, when I left, was almost on Igloo Creek, one-third mile from the bear. The yearling was at first present but later disappeared.

The following morning at 5:20 a.m. the bear was still lying where I had seen him the previous evening. Soon he stood up and yawned, and moved down to the creek and out of sight. The cow moose was 200 or 300 yards from him. At 3:30 p.m. she was on Igloo Creek and at 4:15

she was lying down. At 7:15 p.m. she was in almost the same place. The big bear had no doubt disposed of her calf. The small bear had not dared approach, but apparently the big male had moved forward, and the cow had retreated.

Male Grizzly Undaunted

On 17 May 1961 I spent a few hours watching a pair of mated bears along the East Fork River. When first seen the bears were lying down on the open tundra. During the next couple of hours mating maneuvers took place between them. The male also spent some time following a trail, then again turned his attention to the female.

About noon he stopped bothering her. He raised his muzzle, catching an attractive scent. He climbed a bluff at the edge of the river bar and entered a shallow, brushy depression where he fed, obviously on a carcass because he fed in one spot and I could see that he was tugging at something. I think he had followed earlier the scent left by a fox or wolverine carrying away part of the carcass. Now at last he was at the source. A cow moose now appeared, hackles up as she approached the hollow where the male was feeding, but after a brief look at the bear she trotted away for a quarter of a mile. She repeated this performance twice more, and after the last time, moved up the river for one-half mile. The bear took little note of her and after her last retreat, he apparently had finished the remains, for he moved down to the river bar. Later the cow returned to the hollow, circled it, and finding the bear gone, entered the hollow, where she moved about, nosing it thoroughly. After 15 minutes, she left, but returned three more times to investigate, no doubt hoping to find her calf. Once when she approached, a fox ran out of the hollow. The female bear remained out on the river bar digging roots; apparently she was unaware of the carcass, perhaps being too far out on the river bar to catch any scent. The cow moose was seen 2 days later as she walked a half mile to the hollow where her calf had been eaten. This episode showed that a large, male grizzly could hold his ground in the presence of an angry cow moose.

Cow and Her Yearling Move Away From a Bear

On 18 May 1961 I saw a fairly large grizzly gallop easily across the tundra and enter an extensive patch of willows. A cow and her yearling came out of the willows near where the bear had entered. They trotted away, apparently not frightened but disliking the proximity of the bear, especially in view of his hurried arrival. As I moved nearer, I saw the bear near the edge of the willows feeding on a caribou calf. I did not observe enough to ascertain the circumstances concerning the calf's death, but it seems certain that the cow moose had no calf because she still was tolerating her yearling and gave no indication that she was leaving a calf. Even when no spring calf is involved, it seems that cow moose tend to avoid bears.

Cow and Calf Leave Vicinity of Bear Family

On 19 July 1963 a mother bear and her two 2-year-old cubs were moving north near the base of Cathedral Mountain. A cow moose followed by a calf trotted away about 300 yards ahead of the bears. The moose had left a rendezvous in the willows not far from where the mother bear was waiting for her cubs to leave a fox den. The three bears shifted into a slow lope and trailed the moose. Four sheep that were farther up the slope saw the bears and climbed a little higher, probably not worried so much by the presence of the bears as by their loping gait. The moose disappeared over a rise and the three bears turned, after traveling about 300 yards, and crossed Igloo Creek. The mother bear did not seem too anxious to overtake the moose. The moose only seemed to serve as an excuse for a playful romp, a sort of make-believe chase.

Cow and Calf in Late Summer, Unafraid of Small Bear

On 30 August 1959 I saw a cow moose with a calf on a dense willow–aspen slope below me. Before she and her calf trotted a few yards and went out of my sight she was watching intently something above her. A few moments later I saw a whitish object in the aspens, a cream-colored grizzly, about 4 years old, with brown face and dark legs. Soon the cow moose and her calf returned to where they had been, and after watching the bear a few minutes, browsed in the willows. A little later she and the calf lay down. The bear continued to feed on the slope above her. The moose may have noted its small size and regained her composure, or she may have recognized the bear as one she had seen often and did not fear.

Belligerent Moose Attacks Bear

On 14 May 1961 I stopped near Hogan Creek to watch a medium-sized bear walking west across a hillside of scattered spruces. Ahead of the bear I caught a glimpse of a cow moose, walking toward the bear. Her ears were cocked forward; I guessed she was bent on intercepting the bear, and I was right because a little later I saw the bear making a dodging, scrambling effort to escape, the angry cow galloping close upon his heels and about to strike him with every jump. The bear made a sharp turn as he disappeared from my view and the cow followed, but did not turn quite as sharply. I did not see the bear again, but the cow returned, trotting briskly down a ridge to the road and into the tall willow brush. Once she stopped with ears cocked as though looking for the bear. Soon she climbed the slope, obviously still agitated. A short time after disappearing into the spruces, she reappeared on the open tundra, walking slowly, followed by a tiny calf still unsteady on its feet.

Cow Moose Chases Grizzly

On 1 June 1961 at 11:30 a.m. I stopped to watch a medium-sized bear, a female or young male, as it stood on an open knoll gazing down at a small patch of tall willow brush in a depression. Five minutes later the bear walked down the slope and entered the willow patch. A few moments later a cow moose rushed out from the willows on the lower side and the bear emerged from the point where he had entered. This behavior suggested that the bear had startled the moose, and the noise made by the moose in dashing out of the willows had startled the bear. The bear lay down on the slope, and the moose maneuvered slowly on her side of the willows with ears cocked toward the bear.

Later, the bear walked to the edge of the willow patch and lay down where the willows were less dense. The cow walked slowly toward the bear and as she neared him made a determined dash, causing him to retreat rapidly up the slope. The moose then walked into the heart of the willow patch, where for a time I occasionally could see her as she reached high to browse on willow twigs. An hour later the situation had not changed except that the moose could no longer be seen; apparently it was lying down. The bear moved 200 yards up the slope and lay down for over 2 hours. After that, he fed on the new growth of grass (*Calamagrostis*). When I left the area at 5:30 p.m., the bear was feeding on roots. A few days later I examined the willow patch for traces of calf remains but found none.

A Yearling and Its Mother Flee from Three Moose

On 9 September 1961 on Sable Pass I watched a dark mother bear and her yearling cub feeding on berries on the far side of a creek bottom where tall willow herbs grew. I heard a grunt in the willow patch and the two bears heard it also. They stopped feeding to watch, part of the time standing erect on hind legs. Soon a cow, calf, and a young bull moose emerged from the heavy growth of willows and moved a little closer to the bears, obviously unaware of them. The cub was getting nervous, moved to the far side of its mother, then retreated 10 yards, and stood erect to watch the moose. Suddenly, the cub was overcome with fear and fled, and the mother followed finding it difficult to keep up with her cub. After galloping a quarter of a mile, the mother bear halted and the cub, still ahead of her, stopped also, but he still was very anxious. The mother probably would have moved aside a short distance if the cub had not dashed away.

Bear Family Avoids Bull Moose

On 7 September 1962 I watched a bull moose thrashing vigorously at willow brush with his antlers. A mother bear and her two yearlings, who were foraging in a ravine 75 yards below the bull and unaware of his

Fig. 48. An old bull moose at the beginning of the rut, well able to take care of himself except in deep crusted snow in the spring of the year.

presence, heard the noise and became wary. After listening for a moment, the mother bear led the way down the slope at a steady walk and crossed a little creek a quarter of a mile away before resuming her foraging (Fig. 48).

Bull Moose Frightens Two Yearling Cubs

On 25 September 1963, when bulls were searching for cows, I saw a mother grizzly and her two cubs cross Igloo Creek and climb onto a bench. The mother in feeding soon went over a rise, but the cubs remained in view, feeding on berries. Soon I saw a bull moose walking rapidly on the creek bottom toward the bears. He climbed the slope and came close to the cubs, who, on seeing him, hurried away in the direction their mother had taken. The bull apparently had seen or heard the cubs in the distance and mistaken them for cows. Realizing his mistake, he moved slowly along a contour, grunting intermittently.

Cow Chases Two Bears

On 3 June 1964 a cow moose was standing on a low pass just south of Cathedral Mountain, gazing fixedly toward the Teklanika River. Nothing was in sight so it appeared that she had a scent of something that disturbed her. For 5 minutes she stood watching before she resumed browsing. A few moments later, two bears came over a rise from the direction she had been watching and pursued a course which would take them to one side of her. They appeared to be 2-year-old cubs that I had seen with a female 3 days earlier. As they approached, they stood erect on hind legs three times to look around. They behaved as though they were looking for something, perhaps their mother. They saw the cow moose when they were about 100 yards to one side of her, and loped exuberantly forward. I guessed that they thought the bulky form in the willows was their mother. The cow moose saw them coming and charged. The larger of the two bears veered northward, with the cow moose in pursuit. The cow struck at the bear with a forefoot, but the blow fell just short of his rear. The cow then stopped and, seeing the second bear circling, dashed after it, causing it to flee in the opposite direction. She halted only when she came to a snowfield. The first bear was now circling back to join his companion and the moose chased him again, almost overtaking him before she stopped. After the bears disappeared over the nearby slopes, the moose returned to where she had been, and a small calf appeared out of the brush to nurse. (I learned later that there were two calves.) The cow unhesitatingly had charged two grizzlies. Would she have been as fearless if the bears had been larger, and would a large bear have run away? Possibly a large male would have stood his ground, but in the face of a charge he may have been at a psychological disadvantage and retreated; in addition, the cow might have been less bold if a large male bear had been present.

Cow Moose Chases Lone Bear

On 5 June 1965 about 7 a.m. I saw a cow moose on the edge of an extensive patch of willow brush, and about 100 yards from her a blond, medium-sized bear lying on his stomach, head on paws, watching the cow. Two days before, I had seen the cow at this location with a small calf, and one day before I had seen him a mile from where he now lay.

The cow watched the bear for a few minutes then walked along the edge of the willows. She appeared to be checking on her calf. She turned and walked toward the watching bear, her ears cocked toward him. He saw her approaching and retreated at a walk, and when she started trotting after him, he galloped down the gentle slope. She chased after him, stopping two or three times, and the bear stopped also. After each stop, she would gallop toward him. This continued for a quarter of a mile before she quit the chase and returned to the starting point. The bear continued traveling toward a band of caribou, which he chased briefly. I left the moose standing where I had first seen her that morning. When I returned a few hours later, she was standing in the open as before. Soon she walked eastward, away from the willow patch, and

Fig. 49. A cow moose may put a bear to flight, especially if the bear is not a large adult.

continued for about a mile; then she turned and came back, trotting part of the time. On her return she entered an isolated willow thicket and rested for 2 hours. There was no calf with her, and I suspect that she had lost it, because the following day I saw her leave the area and travel westward for a mile, apparently leaving the area. Perhaps the bear had come upon her calf while she was feeding a short distance away, and killed it. But the calf may have died at birth or been killed by another bear or a wolf, and the young bear may have wanted to feed on what remained of the carcass (Fig. 49).

Moose Chases Grizzly and Wolf

On 27 May, my first day out in the park in 1967, from the road near Hogan Creek, I saw a cow moose and her two recently born calves resting high on a gentle, treeless slope of Primrose Ridge. Near the base of the ridge, a short distance above the road, the carcass of a bull moose was attracting grizzlies and other meat eaters. A grizzly mother and her two 2-year-old cubs were resting on a snow patch 150 yards above the carcass on which they had gorged.

While I watched from a vantage point, two photographers began climbing toward the bears, but before they came in sight, the mother bear retreated up the slope at a slow, deliberate walk, and the two large cubs followed. The photographers soon returned to the road.

The cow moose became aware of the approaching bears when they were a quarter of a mile down the slope. Standing like a statue, big ears cocked forward, she watched. The bears appeared to be unaware of her dark form silhouetted above them, but the mother bear swung slightly to the left as she neared the moose, enough to bypass a hundred yards to one side. Not until then did the bears show any indication of awareness. They stopped, turned their heads for a momentary look, then continued on their way and lay down 300 or 400 yards from the moose, who also soon relaxed and lay down. I learned from the photographers that they had seen this moose with her calves in the same spot the previous day and had watched her chase the three bears as they came near her.

The following morning at 4:45 a.m. I discovered the mother bear and her two cubs high on Primrose Ridge, walking down the slope and passing 75 yards to one side of the cow moose, while she stood watching them. The bears were walking rapidly as though hungry and in a hurry to arrive at the carcass. At 5:10 a.m. the bear family approached the carcass and remained there for about 2½ hours, then moved and lay down on the far side of the creek.

At 9 o'clock the mother bear nursed her cubs and at 10:50 a.m. the family walked up the slope of Primrose Ridge again, at first following the draw they were in where the winter snow lay deep. In half an hour the bears had reached the spot where the cow moose and her two calves had sojourned for at least 2 days. The moose had left the spot soon after the bear family had passed by in the early morning and had gone 400 yards with her calves before stopping. Later, she went over the horizon, how far I did not know. The bears examined the spot where the moose had been briefly and then followed the moose trail at a brisk walk. Soon the bears started loping, the trail apparently making them eager, but before the bears reached the skyline, the cow moose appeared charging the bears, who fanned out and galloped away to avoid hooves. The cow stopped for a moment, then charged again, chasing the bears in a semicircle to some cliffs behind which they disappeared. The cow returned to the spot where she had appeared on the skyline and stood alert, watching in another direction.

Later, I saw a grey wolf approach at a trot within 25 yards of the cow. When she took a few threatening steps, the wolf retreated 10 yards. This was repeated five times—only a slight movement was sufficient to cause the wolf to retreat. The wolf then lay down 40 yards away, but in 5 minutes it trotted off. I left the scene at 2 p.m., but later in the afternoon it was reported that the moose chased the bears and the wolf again.

My observations on grizzly–moose relationships indicate that a cow moose can and will protect her young against bear attack, at least if the bear is not too big. There is some indication that a full-sized, male grizzly may be difficult for the moose to chase away.

It is obvious that the grizzly occasionally captures a calf and that the grizzly is a balancing factor affecting numbers of moose.

Grizzly–Dall Sheep Relationships

The sheep hills are an integral part of the grizzly's home. He uses the slopes for much of his ground-squirrel hunting, root-digging, berrying, denning, and traveling. Because they use the hills jointly, he and the sheep are well acquainted. Each has evaluated the physical prowess of the other and weighed it against his own. The grizzly's knowledge makes it unnecessary for him to squander time and effort in a futile pursuit of sheep if he lacks the advantage, which he generally does. Sheep also are able to judge well the degree of their vulnerability when meeting grizzlies, and so are spared becoming unduly apprehensive. Serious encounters are uncommon. Only occasionally does a grizzly have the opportunity to capture a lamb, and even more rarely, an adult. Because the grizzly is a potential enemy and the sheep potential prey, each plays a part in the life of the other (Fig. 50).

The sheep enter the bear's diet chiefly as carrion. The extensive wanderings of grizzlies in early spring no doubt take them occasionally to the remains of a winter kill sufficiently intact to furnish some food.

Although bears, and sheep to some extent, feed on berries, competition is insignificant because in most years enough berries are available for all unless a drastic berry failure occurs, and then there is little for anyone. On one occasion I did see what might be termed some direct local competition for berries, when 131 migrating sheep crossing the Toklat River stopped in one of the grizzlies' favorite buffaloberry patches and fed extensively.

Sheep Alert to Presence of Grizzly

The behavior of sheep may call one's attention occasionally to the presence of a bear, for when a bear is sighted sheep often stop feeding and watch. If he is distant, they may resume feeding; if nearby, they may watch until he has passed or make a precautionary move up the slope and then resume feeding. A running bear causes more concern than a feeding bear, which is likely to receive only perfunctory attention. On 6 September 1963 I saw 70 ewes and lambs move up into cliffs when they saw a mother bear followed by a spring cub galloping and hunting ground squirrels on a contour far below them. On other occasions I have observed sheep reacting similarly upon seeing a bear loping away from a man.

I frequently have observed sheep close to bears without showing much concern. In 1953, four of us climbed Sable Mountain to look for White-tailed Ptarmigan, to observe the fall coloring of herbaceous cinquefoil and other herbs, and to enjoy this unspoiled, wild country. We saw four sheep following a ridge in our direction, and, thinking it would be interesting to watch them pass close by, we remained hidden behind a light, rocky prominence. While we waited, a grizzly appeared over a

Fig. 50. Dall sheep—three ewes and three yearlings—not much worried about bears; but under certain rare circumstances even an adult sheep may be captured by a bear.

side ridge from the opposite direction, traveling our way. When the grizzly and the sheep were opposite each other, a little above us, the grizzly changed course slightly, veering toward the sheep which moved to one side and a little higher on the ridge, watching while the grizzly continued traveling, passing between us and the sheep. The sheep appeared to feel no danger, and the bear apparently sized up their situation in the same way.

On 25 May 1955 I watched a female and two 2-year-old bears climb a slope near Savage Canyon. When they neared a band of 33 rams, the mother made a short run toward them. The rams fled upward about 100 yards, then walked slowly a little farther, and stood watching the family pass over the ridge a short distance from them. In rough country the sheep are aware of their security.

On 24 May 1963 I watched a ewe and lamb move across a rather gentle slope on Cathedral Mountain and then discover that they were below a mother bear and two 2-year-olds who were busy digging roots. The ewe recognized her position as vulnerable and galloped rapidly across the contour below the bears and up the slope to the other side. Here, although still not far from the bears, she felt safe.

The following day I saw on the same contour a young ram approach the three bears who were digging on the same gentle slope. He stopped to watch the bears when about 75 yards from them, then turned and moved slowly up the slope to one side, soon confident in his safety, for rugged cliffs were nearby.

Migrating Sheep and Grizzlies

When sheep, in migration, pass across long, gentle stretches of terrain such as valleys and river bars, they are vulnerable if they are discovered by a wolf, and probably are somewhat vulnerable to attack by bears. There have been three or four incidents described to me in which migrating sheep in the Toklat River area were hard-pressed to escape a bear. On one occasion a bear cut a ewe off from the main band and nearly overtook her as she crossed a spring snowfield. Fortunately for the ewe, she managed to cross the snow and gain nearby slopes where she had the advantage (Fig. 51).

On 8 June 1961 I expected to see an exciting incident when I observed two ewes, each with a lamb, crossing the wide Toklat River bar on their way to Divide Mountain. They were headed straight for a big male grizzly and his mate who were on the bar digging roots. But the sheep veered to one side of the bears, which they could not see because of scattered willow brush, and crossed without incident. If they had met the bears, it is possible they would have had some difficulty escaping.

Behavior cannot always be predicted. On 22 September 1961 I watched a young bear, perhaps a 3-year-old, traveling down the Toklat River bar near the forks. As he proceeded, he approached a group of about 60 ewes and lambs that had scattered widely over the river bar, feeding as they crossed. Although some of the sheep were watching the bear, when he was 200 yards away they did not take flight as I expected. The bear must have seen the sheep, though from his actions I was not sure that he had. Possibly this young bear was not perceptive enough to realize that the sheep were vulnerable, and so paid no more attention to them than if they were in cliffs. He went about his business as though they did not exist, turned aside to sniff a squirrel hole, and a little later made a right-angle turn; in a few minutes he was lying on his back scratching himself. The lack of concern that the sheep showed was harder to explain. They walked down the bar away from the bear and continued for 200 or 300 yards before stopping to feed again, but I expected more concern. If the bear had given chase, I think the sheep could have crossed the bar to safety but they did not seem to take their usual precaution to insure maintaining control over the situation with room to spare.

Bear Captures Newborn Lamb

On 21 May 1949 along Igloo Creek a group of men were working on the road and saw four ewes and a newborn lamb in a draw almost at the

Fig. 51. A band of Dall sheep, in their migration across a valley, could not resist stopping in a wet meadow to feed on horsetail and grasses. In this situation they are somewhat vulnerable should a bear happen to come upon them.

bottom of the creek. One of the men approached the sheep to take a picture. When he was ready to take the picture, a bear galloped down the slope toward the sheep and the photographer. The photographer escaped with long strides across a snowdrift, but the newborn lamb, after a chase of a few yards, was captured quickly and devoured. The photographer thought the bear was charging him and was diverted by the sheep at the last moment; the spectators believed that the photographer was incidental and that the bear had the sheep in mind all the way down the slope.

On the following morning when I came upon the scene, the bear was some distance up the same slope digging roots. The mother sheep was lying near the spot where she had lost her lamb. In about 20 minutes the bear walked slowly down the slope toward the ewe, who now stood up and picked her way along a low rim bordering Igloo Creek. The grizzly followed her trail and stopped now and then to paw out a chunk of sod to get at peavine roots. His interest in the ewe persisted, with good reason. The ewe's rear was bloody, and about 8 inches of afterbirth still dangled. No doubt a little blood was in her trail, enough to encourage the bear. Shortly before the bear arrived at the fox's den where the ewe was now lying, she stood up and picked her way down the rocks at the point of the ridge. The bear came to the point directly above her, so close it seemed he might have rushed her successfully, but he came slowly down one side, and by that time the ewe had crossed a sidestream by clambering through deep snow. She was soon on a long ridge leading up Igloo Mountain, and the bear, after following for some distance on the ridge, gave up and returned to digging roots.

The evidence indicates that this incident developed from special circumstances. Normally, the lamb would have been born high in the cliffs where it would have been well protected. The prolonged retention of the afterbirth suggests an abnormal birth, and a weak lamb perhaps, causing the ewe to drop the lamb in the unprotected spot near the stream.

A Fortuitous Incident

On 27 May 1950, I just missed seeing a grizzly capture a lamb. I stopped to classify a group of ewes and lambs resting on a sharp, grassy spur ridge on the south slope of Igloo Mountain. A short distance below this group, two additional ewes and a lamb were grazing. After classifying the sheep, I drove down the road and stopped for a look at the other side of the ridge to see whether any sheep were resting beyond my first view of them. While I watched, the group I had classified galloped into view and crossed a shallow ravine. This seemed rather unusual, for I had seen nothing to alarm them. I thought they might be frolicking. I drove back up the road, saw the two ewes and lamb still grazing, and decided they had been frolicking. About 5 minutes after I left the scene, a road worker came by, heard a mother sheep bleating, and saw a bear feeding on a lamb out on the open slope. The bear had escaped my observation and also that of the three undisturbed, grazing sheep. The uneven conformation and sharp contours of the slope below the ridge apparently made it possible for the bear to approach within a few yards of the sheep without being seen by them, and the lay of the land had kept the bear out of my view. I saw the mother sheep near the scene the following day, and on the next day saw her traveling in high cliffs, occasionally stopping to look down at the ridge where her lamb had been lost.

Trailing Lamb Captured

On 7 June 1963 a large group of ewes and lambs was feeding near the north end of Cathedral Mountain. I spent about 2 hours photographing them as they crossed two spur ridges. When I came back over the second spur, I saw a mother grizzly with two spring cubs on the adjoining ridge where I had first started photographing. Only a narrow, shallow draw was between us, so I angled down the slope to get out of her way. When I looked again, she was galloping at top speed down the slope directly toward me. I was dumbfounded, but fortunately she was not after me, but soon overtook a very young lamb. A ewe, a yearling, and this lamb had crossed Igloo Creek earlier, and naturalist Verde Watson had seen them closeup. He said the lamb seemed quite young and weak because it lagged behind the ewe and yearling. When the bear started the chase, the lamb tried to angle up the slope but then, perhaps because of weakness, turned and ran directly down the shallow draw where the bear captured it quickly. The bear fed for about 10 minutes at the spot where the capture was made, and then carried the carcass 15 yards up the slope and finished it (Fig. 52).

The following morning she left her bed soon after 5 o'clock and dug roots nearby for half an hour. She then climbed over the top of the ridge and went out of sight into a basin. From this basin emerged a canyon where I earlier had seen the big herd of ewes and lambs, scattered widely as they fed. To watch what happened when the bears arrived, I returned to the herd of sheep. Soon, a dozen lambs up the slope apparently saw the bear, for they watched in her direction and then started moving away slowly. Some ewes lower down behaved as though they too had discovered the bears. Then I saw the mother bear galloping rapidly on a contour toward some sheep bunched up ahead of her at a slightly higher elevation. The bear rounded a shoulder, galloped up the steep slope a short distance, then stopped to look around. She seemed aware that opportunities for capturing a lamb might exist. Turning, she galloped up the slope a short distance toward the main group, then stopped, realizing it was hopeless. The sheep soon returned to normal activity as the mother

Fig. 52. Not long after this picture was taken a bear captured a straggling lamb that was following its mother up a slope. It was an accidental meeting, and the lamb apparently was weak.

bear moved to one side and dug roots. This bear, because of its good luck the previous day, probably was more optimistic than usual.

He "Had It Made" and Didn't Know It

On 27 August 1963 I watched a band of 73 ewes and lambs feeding on a low, gentle slope of Sable Mountain. I noted that one of the lambs had difficulty lying down and walked stiffly. It was obviously ailing. Soon, I saw some of the sheep that had been feeding to the east of the main group run up the slope and stop to look eastward. Across a narrow draw, about on the same level as some of the sheep, I saw a dark grizzly walking steadily along a contour in the direction of the sheep. I estimated that he was 3 or 4 years old. Because of a short, new coat he seemed especially rangy. As he emerged from the draw, the sheep that had seen him angled westward and higher up the slope. When the bear was about 300 yards from the main herd and the sick lamb, now lying 30 yards below the others, he stopped to investigate some squirrel holes, but he soon moved forward again. When the main herd saw the bear, they galloped en masse westward and upward. The bear, upon seeing the general flight, loped forward. He was slightly above the sick lamb and perhaps 150 yards from it. The lamb got to its feet and, spurred by fright, galloped away rapidly, at first on the contour, then, being weak, turned down the slope. In full flight he collapsed and lay in an inert heap. Upon seeing the lamb collapse, the bear stopped and seemed puzzled. After watching for perhaps a minute, he did not hurry to the lamb, but turned slowly as though questioning what he should do, then loped back over his trail to where I first saw him, and walked steadily down the hill out of my view. Apparently, he was an inexperienced bear overwhelmed by the flight of the many sheep and the unusual and dramatic behavior of the sick lamb. He played it safe and retreated. It was a case of the elimination of the weak except that the inexperienced bear failed to play his role.

Bear Captures Yearling Sheep

About noon on 23 May 1964 I saw a fairly large blond bear arrive at a snow-free strip on the crest of a ridge. Tracks behind the bear showed that he had been traveling on a contour across snowfields that lay between low parallel ridges leading down to Igloo Creek. The bear had stopped to look around before starting to walk down the ridge he was on. Soon he stopped again and gazed steadily at an angle toward the creek below. Following the direction of his gaze, I saw a cow moose 250 yards from him, moving along the edge of the bench. Below the bench the slope dropped off some 30 yards to the creek bed, which was still covered with winter snow and ice. Following 50 yards behind the moose was a yearling sheep—an unusual combination. Their tracks in the snow showed that they had moved down the ridge, starting from an area near the spot where the bear was standing. The moose moved diagonally down the slope; the yearling sheep traveled along the bench, toward the moose. Evidently, they had both seen the bear coming along the contour and had retreated to avoid him. A ewe was still up on the ridge out of sight of the bear and only a short distance leeward from him, and neither was aware of the other.

The bear now loped easily down the slope toward the moose and the sheep, who were moving along the bench and out of my view. (I did not go forward at once for fear of intruding on the situation.) Apparently, the bear was approaching them at this easy lope to investigate rather than with an immediate expectation of a capture. Halfway down the slope the bear stopped briefly to look, then continued as before, loping through the deep snow, until he too was out of my view. In a few minutes I had moved forward far enough for a good view. The moose had disappeared down the valley. I saw the bear dragging the yearling off the creek bottom and up the steep slope. As the bear moved up the slope, the carcass dragged to one side or the other and sometimes between his front legs.

Apparently, as the sheep started down a cliffy portion of the slope, it collapsed, rolled, and dropped off a perpendicular part of the cliff. Patches of hair clung to the brush where the sheep rolled down the slope. A bloody depression in the snow showed where the sheep struck the creek bottom and the bear, coming down to one side, killed the sheep. If the sheep had been in good health, it probably would not have retreated to the creek, or it would have been able to cross the creek and escape up the far slope.

When the bear reached the top of the bench, he dropped the carcass and walked a few steps to the edge of the bench to survey the creek bottom. Returning to the carcass he began to dine, first biting mouthfuls of hair from the hide to get at a hind quarter. Much of the time he sat on his haunches as he fed.

After eating for 45 minutes, he started pawing debris and sod toward the carcass and over it, using slow, deliberate strokes with one forepaw at a time. Two magpies discovered these activities and came as close as they dared for tidbits. In an hour the carcass was well covered with a mound of debris and the bear was lying beside his treasure. I left and returned a few hours later (6 p.m.) to find him lying on top of the cache which had been heaped up higher during my absence.

The following day, the 24th, a storm and drifting snow prevented me from returning to look for the bear. On the 25th I did not see the bear but saw his tracks and fresh droppings containing sheep remains.

On the afternoon of the 26th I found the bear picking away at the sheep skeleton, feeding lackadaisically for an hour. Two magpies, old friends by now, paid him a brief visit. The bear walked slowly away for

20 yards and lay down for 3 hours. During these 3 hours he rested in a variety of positions, reclining on back, stomach, and both sides, with variations of each. For a time, on his back, he rested like a female nursing young, with head raised so as to see nursing cubs if there had been any. Toward the end he became restless, changing his position often. As he departed, he ate a few mouthsful of snow, and looked around as though wondering what he should do. Then he waded 100 yards or more northward through snow, made a wide arc, retraced his steps in the arc, and walked hurriedly to a bare spot on a ridge where two magpies were picking at a bone. He chewed on some scattered pieces on the bare spot, then out in the snow he uncovered what appeared to be the carcass of a calf moose. Suddenly he became alert, stepped forward a short distance, stopped, and peered into a canyon (out of my view) for 5 minutes. Possibly the mother of the calf was in the canyon. He fed for 1½ hours, pulling small pieces of meat and tendons from the bones.

The next day (27 May) I saw the bear approach the sheep carcass from above, wading in deep snow. When in view of the cache, he stood watching for 4 minutes, probably to learn if another bear was in the locality.

At the cache he lay on his stomach and fed, pulling meat and sinew from the bones. Each tough little morsel was given 30 or 35 vigorous chews. After a time, he was pulling tidbits from pieces of hide. He pawed at the debris he had used to cover the carcass as he searched for pieces too insignificant to notice in earlier days of plenty. After 1 hour and 10 minutes, he lay down on the debris. During the next hour, the bear moved off twice to eat snow. After another half hour of resting, he yawned three or four times with tongue stretched out and forefeet forward. He sat for 8 minutes testing the air. It was a warm, sunny, quiet day. After departing he soon returned to the cache, contemplated it, and yawned some more. Soon he was chewing bones and pulling loose tough sinews, chewing each bite 70 or more times—the remnants were becoming tougher. He covered the remains with debris and lay down. The yearling carcass had been a point of interest for 5 days.

Evidence Indicated a Bear Captured Newborn Lamb

On 30 May 1961 a ewe gave birth to a lamb on a gentle slope at Polychrome Pass. In my notes for that day I wrote that the lamb was on the gentle slope and would be easy prey for a bear or wolf. On my way home I saw a dark bear about half a mile from the mother and lamb. It was climbing a ridge, headed in their general direction. I wondered if the bear would wander in sight of the ewe and lamb, but the country was somewhat broken and the chances seemed good that the bear would not come upon the sheep.

I returned to the sheep in the morning and saw the ewe feeding about 60 yards from the spot where I had last seen her and her lamb. Two hundred yards higher up on the ridge was the bear I had seen the day before. The bear was resting but soon moved over the ridge out of view. The ewe walked steadily to the point where the bear had disappeared and stayed there an hour. Later, she returned to the birth site. The following day the ewe was still in the area, searching for her lamb. Circumstantial evidence indicated that the bear had eaten the lamb.

Unusual Capture of an Older Sheep

On 23 July 1964 a photographer who had been in the sheep hills during the day told me about an unusual incident. The father of the photographer watched from a distance while the boy stalked some sheep for a picture. A bear that earlier had been seen moving up the slope suddenly came upon one of the sheep that the boy was photographing, and pounced on it. The bear had not stalked the sheep but came upon it accidentally. The sheep had 5-inch horns, either a ewe or a young ram.

My observations indicate that a grizzly occasionally captures a weak sheep or a very young lamb. Most captures seem to result from chance encounters, in the same way that sheep carrion occasionally enters the bear's diet. Sheep, even lambs, are not actively sought as caribou calves sometimes are.

7
Grizzlies and Rodents

Grizzly–Ground Squirrel Relationships

The Arctic ground squirrel is common over all open country in the park, from the lowlands to the tops of the sheep hills. It is plentiful year after year, apparently undergoing no marked cyclic changes. Perhaps this is due to the steady, heavy pressure on the species by bears, foxes, wolves, and Golden Eagles, a pressure that does not permit overpopulation and resulting die-offs (Fig. 53).

As a bear travels or moves along feeding on vegetation, he may surprise a squirrel and capture it before it can escape into a burrow. If the squirrel does manage to reach a burrow, the escape may be temporary for the bear will excavate, nearly always with success. When the bear happens to encounter a set of squirrel holes, he gives them a routine inspection with his nose, and if the scent indicates that the squirrel is at home, the bear begins to dig. Usually squirrels are dug out when the bear happens to encounter a promising set of burrows. But at times, chiefly in late summer and autumn, squirrels are hunted systematically. I have noted more squirrel hunting at this time than in early summer.

Considerable excavating usually is necessary before a squirrel is captured. The bear may dig into three or four entrances in a set of burrows and in at least one he may dig so deep that his shoulders are hidden from view. While excavating at one burrow entrance, he tries to keep a sharp watch on the other entrances, for he knows that the squirrel may emerge from any one of them. He is not always successful in capturing his prey; even after laboring for as long as half an hour he may give up the job as hopeless. Occasionally, a squirrel escapes from one set of holes and vanishes into another set a short distance away, and the bear must then begin to dig anew. The bear's luck varies. He may be successful in a series of diggings, or he may try several times without results.

Fig. 53. Bears often are seen excavating ground squirrels, which are uniformly abundant year after year.

In late May and early June bears may be especially fortunate in their digging and come upon a nest of young squirrels, but more often the reward is a single squirrel (Fig. 54).

Jarring of Sod Causes Squirrel to Emerge

In excavating a burrow the grizzly may use a single paw or both paws together. He may pull away loose soil in the burrow for a time, then with both front paws push down and pull on the sod at the edge of an excavation to make it cave in, thus making a bigger opening and more room for digging. This pushing action, given joltingly, frequently jars the earth enough to cause the squirrel to run from one of the holes and be caught. At times this jarring action is performed only to scare out the squirrel. As he jars, the bear keeps a sharp lookout for the squirrel, knowing that it may emerge. On a few occasions I have seen a grizzly jump with the forefeet, a kind of pouncing movement, several times during excavation of a ground squirrel burrow.

On 27 August 1963 I watched a mother bear digging for a squirrel on Sable Pass. She had not dug far when she began jarring the sod on the upper edge of her excavation. No effort to loosen the sod was made. She obviously was trying to frighten the squirrel enough to make it emerge, which it did after she had struck the sod with both paws five or six times. After eating the squirrel, the mother and one of her two yearlings moved away and out of sight of the second yearling which was left behind digging at another set of squirrel holes. The yearling dug for several minutes and finally captured a squirrel—one of the few times I have seen a cub do so.

Mother Does not Share Squirrels With Cubs

The cubs usually wait docilely for the mother to eat her squirrel, but not always. On 23 July 1959, when a mother captured a squirrel that had emerged from a set of holes where she had been digging, one of her two yearlings growled and cried while she consumed the squirrel. This complaining continued for a minute or so after she had finished eating.

In August 1969 I observed a mother with one spring cub hunt ground squirrels near Tattler Creek for several days. The female chased one ground squirrel to a burrow and began to excavate. The squirrel slipped away to another burrow and the female followed. As the mother resumed her digging, her cub chased another ground squirrel to the first burrow. Apparently this burrow was now plugged because the cub caught the squirrel. This rare moment of success was short-lived; the mother ran over, snatched the squirrel, and ate it as the cub watched somewhat disconsolately.

Fig. 54. A young bear looking for a ground squirrel.

About One Hour Expended to Catch a Squirrel

On 21 September 1950 I discovered a mother bear almost buried in an excavation she was making trying to dig out a ground squirrel. She had been digging for some time, judging from the depth of the hole. Her two spring cubs, in the meantime, had wandered almost half a mile away. Even after I discovered these bears, the mother continued working at the excavation for 45 minutes. After digging a while, she would put her nose to the bottom of the hole to test the squirrel scent, then raise her head above the surface, mouth open and panting, and look to see if the intended prey were trying to escape from another entrance. A few times she made four or five jumps up the slope to make sure the squirrel was not escaping. She would stand on hind legs and look around before returning to the digging. She kept caving in the sod at the edge of the excavation, pulling loose large pieces, and pawing them out from the bottom of the hole. The sod and loose dirt flew between her hind legs or off to either side. Sometimes a chunk of earth or a rock would make a noise loud enough to startle her and she would turn quickly to look. This happened four or five times. After a while she was hidden completely in the hole. Once she came forth to sniff at other entrances and began to dig at one of them, continuing until half-buried before returning to the main excavation. The cubs returned and after resting sat up watching the mother. At last, from her deep hole, she managed to bring forth a squirrel. She bit off small pieces and chomped with a wide open mouth at each one. She chewed five or six pieces before the squirrel was eaten. I have seen bears swallow long, thick pieces of caribou tendon without chewing, perhaps because they were too tough, but squirrels usually are eaten in small pieces with much chewing. The cubs moved a little closer, but just sat and watched the mother dine. Before wandering away, the mother made some final sniffings into the large excavation. This bear had dug for about an hour before making the capture; usually a grizzly will give up before expending this much effort at a squirrel hole. The small result probably did not compensate for the energy expended. This incident is a record for digging time, but on another occasion I watched a bear dig for 40 minutes and he also had begun before I first saw him.

A Squirrel Escapes

On 31 August 1963 I discovered a bear digging for a squirrel that had taken refuge in a burrow leading under a large boulder about as high as the top of her back. Her eagerness indicated that the squirrel's scent was strong. Digging mostly from the upper side of the boulder, she would dig for a few seconds, then crane her neck to see if the squirrel was emerging on the other side. Sometimes she would try to look over the boulder and continue to dig, rather ineffectively, with one paw. After about 15 minutes, while the bear was intent on digging, the squirrel sped from the other side. When the bear looked up, the squirrel had crossed a draw and was part way up the other side. The bear galloped after the squirrel but it escaped into another set of holes. The bear dug briefly at five entrances to this hole then concentrated on one, working until only her hind quarters showed above the excavation. During this time I heard a loud bawling, and 75 yards off I saw a lone, impatient cub wandering back toward its mother from the other side of the slope. The mother dug for another 15 minutes, then stood looking over the valley, mouth open, before proceeding briefly with her digging. Farther down the slope, she ate a little horsetail, then wandered away into the gathering dusk. The squirrel was safe for one more day.

A Mother Intent on Her Squirrel Hunting

In 1963 the berry crop failed in the high country, so the bears were forced to rely more than usual on other foods. Ground squirrel hunting also seemed more prevalent that fall.

On 8 September 1963 a mother whose spring cub was so crippled that its hind foot was useless, left this offspring far behind when she went hunting. I first saw the cub alone, bawling steadily, and it disappeared into a canyon. Fifteen minutes later the mother, having gone up the canyon, came over a ridge and the cub was soon on her trail, still complaining. The mother galloped along, looking expectantly for ground squirrels. In a patch of willows she dug for some time. Once, she emerged from the willows, looked around for the squirrel, and returned to her digging. Five minutes later she dashed into the open again, this time chasing and capturing the ground squirrel.

Five days later this same mother was observed on Sable Pass about 3 miles farther south. She alternately walked, trotted, and galloped as she moved along on a contour, looking for ground squirrels. The cub was 200 yards behind, limping along on three legs. The mother showed no concern until she was about one-half mile ahead of the cub. Then she lay down on a knoll facing the direction from which she had come, but before the cub reached her, she was on her way again, once running 30 yards after a ground squirrel but missing it. The cub finally reached its mother and they both disappeared for 25 minutes behind a knoll. The mother reappeared and sniffed around for half an hour, digging at one burrow and exploring two others. The cub rested and licked his injured foot. It was an unfortunate time to be injured because the mother was traveling more than usual in her hunt for squirrels. She made several short runs with ears cocked. In one set of holes she dug deeply at four entrances, moving from one to the other, poking her nose into each one. While she was busy at this den, a second squirrel emerged some 30 yards below and made short runs up the slope until it glimpsed her, whereupon the squirrel scurried a few yards away and sat erect and perfectly still for 15 minutes. In the meantime, the grizzly captured the squirrel she was after and came a few steps down the slope, but fortunately for the second squirrel, she turned aside. The little animal scurried 15 yards to a hole without being seen.

The mother, still hunting for squirrels, returned to where I had first seen her. Later, the resting cub tried to follow her trail but could not find the beginning because it circled and criss-crossed near the excavated holes. The cub bawled as he searched and was soon circling off to one side, getting nowhere. Finally, he started up the slope and after climbing 200 or 300 yards, found his mother's trail. His crying ceased, for his most pressing worry had been relieved, but soon he was crying again. Eventually, he saw the mother digging for a squirrel, but as he came near, she was traveling again, doubling back over the same route to the base of Sable Mountain. The cub took a brief rest, then followed her trail, again crying bitterly. Later, the mother led the way over a low ridge as she headed for Tattler Creek.

Such eager, prolonged hunting of squirrels is unusual. I have often observed squirrel hunting in the autumn, but none with quite this degree of energy and drive. Moreover, I have never seen a mother show so little concern for her cub. Apparently she was accustomed to having her crippled cub lagging far behind and crying.

In 1969, another mother with one spring cub spent much of its time hunting squirrels in mid-August. Over 8 days I watched her capture seven ground squirrels, and she was alert constantly for opportunities.

A Mother has Unusually Good Luck

On 1 September 1959 I watched the mother of two yearlings dig out a squirrel from a snowy hillside. The mother jumped at the squirrel four or five times before capturing it, sometimes so vigorously that clouds of flying snow almost hid her from view. I watched as she hunted squirrels for 2 hours and 45 minutes, catching nine. The two yearlings followed along, resting while she dug. As usual, they did not share in the catch nor did they expect to do so. The mother's success varied from hole to hole. Once she chased a squirrel into a hole so shallow that excavation took only 2 or 3 minutes. With little digging, she secured two more squirrels but at the next hole she almost buried herself before reaching the squirrel.

A Family is Unsuccessful

One day in late August 1969, near Toklat, I watched for several hours as a mother with two yearlings traveled, occasionally digging for roots or ground squirrels. The female began digging at a ground squirrel burrow in a patch of low willows, and when the squirrel ran from the hole, all three bears pounced after it, at times almost colliding with one another.

Despite their efforts, or perhaps because of their mutual interference, the squirrel escaped to another burrow. The mother and one cub each dug at a different entrance, and the second cub watched, but they soon gave up and moved on.

Ground squirrels recognize the grizzly as an enemy and utter loud, sharp warning signals when one is near. All squirrels in the neighborhood sit erect and join the warning chorus. Frequently, squirrels living at our camp warned us when bears, wolves, etc., were passing. Many false alarms seem quite authentic until one experiences the genuine alarm call, which is unmistakable. Although ground squirrels are only a small part of the bear's nourishment, they do add some meat to his diet and considerable interest to his daily living.

Grizzly–Marmot Relationships

Hoary marmots frequently send forth loud, sharp warning whistles when they discover a bear, but they are rarely excavated from their secure burrows among rocks and cliffs. Just as humans find a marmot occasionally away from the protection of a burrow, I expect bears sometimes come upon one too far from a retreat to escape. On a few occasions I have observed a bear investigate a marmot den but I have never seen a bear trying to dig one out. Marmot remains were seldom found in grizzly scats. However, if a marmot denned away from rocks, excavation by a bear would be a danger. Perhaps the bear is one of the factors causing marmots to live among rocks (Fig. 55).

Grizzly–Mouse Relationships

Seven species of voles and lemmings have been found in the park. In years when one or more of those species are plentiful, bears sometimes feed on them. Mice are a tidbit, and so also are the underground stores of roots and tubers cached by the hay mouse (*Microtus gregalis*). Among the plants represented in the caches are coltsfoot (*Petasites*), bumblebee plant (*Pedicularis*), horsetail (*Equisetum*), knotweed (*Polygonum*), and peavine (*Hedysarum alpinum americanum*), the species the bears dig for constantly.

A cache may contain 2 or 3 quarts or more of roots, sufficient to make the excavation worthwhile. These mouse caches also are known to the Eskimo. Porsild (1953) says that Eskimo ". . . rob the mice caches which they locate by means of a dog specially trained for the purpose." Bears find the root caches (and the mice) with their keen noses, and probably learn quickly that mice favor hummocks for their nests and caches.

In 1955, a year of mouse abundance, bears were observed frequently digging into hummocks to obtain mice or their stored roots and tubers.

Fig. 55. Hoary marmots are present in grizzly habitat but seldom are captured or excavated from their secure burrows in rocky areas.

On 20 June, a mother and her two yearlings, after feeding for a time on the previous year's crop of crowberry and cranberry, moved up into an area of hummocks, and dug out one hummock after another, for over 2 hours. In the same year, on 9 July when green foods were available, a mother and yearling spent much time digging for mice. A few days later, 16 July, although feeding chiefly on green foods, a mother dug into several hummocks. On 11 June 1959, although mice were not especially abundant, I found some root caches exposed and eaten by bears. In 1907 (a good mouse year) Charles Sheldon (1930) observed the grizzly feeding extensively on mice. He describes observations made on 9 October as follows: "The bear, evidently scenting a mouse in a tunnel, would plunge its nose into the snow, its snout ploughing through, often as far as ten feet, until the mouse had gone down into its hole in the ground; then the bear would dig it out and catch it with a paw."

Grizzly–Beaver Relationships

I have seen no evidence of predation on beaver by grizzlies in the park, and I have only two records of grizzlies feeding on beaver carcasses there. One carcass was known to be carrion, and the other probably was also. I have no observations suggesting bear predation on beavers, and it seems unlikely that beaver serve as anything other than an occasional taste of carrion for the bears.

Grizzly–Porcupine Relationships

I have no evidence of grizzlies killing porcupines or vice versa. However, occasionally there is contact and sometimes a grizzly is injured or a porcupine killed, but the latter is rare.

Discreet Behavior

The grizzly usually avoids the porcupine. For example, on one occasion I saw a plodding porcupine approaching a feeding bear. The porcupine stopped a few yards away, the two looked at each other, and the bear watched while the porcupine, maintaining his dignity, turned slowly and detoured past the bear. Alfred Milotte reported a bear watching while a porcupine approached and climbed a tree. Once, I saw a female grizzly watching a porcupine waddling past only a few yards away. A big male standing next to the female did not even deign to look, and the porcupine, for his part, seemed unperturbed by the presence of the bears.

Indiscreet Bears

Occasionally, bears fail to take proper precautions when they are near a porcupine. A bear appeared once near park headquarters with several quills in its face. The bear was seen rubbing its head against a tree,

Fig. 56. Occasionally the curious paw of a grizzly cub encounters quills, and adult grizzlies have been seen carrying quills on nose and face. As a rule the bears seem to keep a safe distance from porcupines.

apparently trying to get relief. At least two other bears were reported with a few quills in their noses (Fig. 56).

On 2 June 1959 I saw a young bear with a crippled front foot that caused him to limp. The track of the normal paw was a little over 5 inches wide, that of the injured foot, about 4 inches. This suggested an old injury. In digging roots only the normal paw was used. Two days later this bear was shot at a campground. The claws on the uninjured foot were worn down, probably due to excessive use in root digging, and the claws of the crippled foot were unusually long from lack of normal use in walking and digging. Examination showed a number of old porcupine quills buried deeply in the injured foot. There was considerable festering in the foot and under the shoulder blade on the same side. The bear was thin, so although the injury was not fatal, it was crippling.

On 27 May 1959 while watching a mother and two 2-year-old cubs at Milepost 28 along the Teklanika River, I noticed that one of the cubs limped. On 7 June I watched the family feeding on berries near Hogan Creek. A few times the lame cub lay on its back to chew at the injured foot. Its jaws bothered it too, for it chomped them, pawed at its face, and shook its head impatiently. I could not see the quills, but the action made it almost certain that this cub had tangled with a porcupine. The following day, when I saw the bears on a distant slope, the cripple was still limping. On 10 June these bears were digging roots on flats near Sanctuary River. The cripple rested much of the time, once for 25 minutes, while the other two fed. When it walked, the injured foot was carried or used lightly. Sometimes it rested on the elbow instead of on the injured paw as it dug roots with the other paw. When I saw these bears on 11 June, the cripple was resting while the other two dug roots. The cripple obviously was not eating as much as its twin, and appeared gaunt. Its condition suggested that it might not survive the coming hibernation period.

On 25 June 1964, a photographer told me that he had seen a blond grizzly "explode" out of a patch of willows and that a few minutes later a porcupine emerged on the opposite side. When I arrived at the scene, I saw the bear on the slope biting at his paw.

Isabelle and Sam Woolcock reported watching a young bear jumping playfully about a porcupine. The bear seemed to understand that the porcupine was not to be touched; nevertheless, there was a chance for an accident. The porcupine climbed a stout willow where I saw it an hour later.

Some cubs may learn about porcupines by observing the behavior of their mothers, and others may learn from experiences. Incidents similar to the one told about a black bear also may occur to grizzlies.

> J. K. MacDonald, of the Hudson's Bay Company, told me that about 1880, when on the Sascateway River, Lake of the Woods region, he saw the body of a black bear that was killed by porcupine quills. Its mouth and lips were full of them, and its head swollen to a frightful size, nearly two feet across. The unfortunate creature could neither open nor shut its mouth; it was found starved to death in a pool full of suckers which it could easily catch, but could not eat (Seton 1929).

Species other than bears sometimes are indiscreet about the porcupine. Foxes have been found injured seriously by quills, and unsophisticated dogs are stuck frequently by quills. In *No Room in the Ark*, Alan Moorehead (1959) writes about superannuated lions: "Often in their extremity old lions pounce on a porcupine, and that leaves them lame with a mass of quills in their paws."

It also is possible that the porcupine contributes to the improvement of the bear's habitat. At a spot near the Toklat River porcupines have killed most of a patch of spruce, and consequently willow brush and horsetail a favorite grizzly food have increased. In June, grizzlies feed extensively on horsetail in this patch of dead spruces. One day I collected 12 fresh bear scats in a few minutes, all containing only horsetail.

Fig. 57. Bears often benefit from animals killed by wolves. A lone bear sometimes is harassed by a group of wolves and then may be glad to retreat.

8
Grizzlies and Carnivores

Grizzly–Wolf Relationships

Both the grizzly and wolf are fond of carrion; consequently the two species renew acquaintanceship occasionally at a carcass. Regardless of who arrives first, the bear generally takes possession and may camp near the carcass for as long as it lasts. The carrion may result from disease, old age, or an accident, but often it results from hunting by wolves who may get a meal or two before the bear is alerted. The bears appear to benefit most from the relationship, primarily because they partake of many wolf kills (Fig. 57).

The following incidents observed in the field illustrate the relationship between the two species.

On one occasion a mother bear and three 2-year-old cubs approached a wolf den from downwind (Murie 1961). The four adult wolves at the den did not notice the bears until they were close, whereupon the wolves dashed out in a vain effort to protect their property. For the hour that the bears were at the den feeding on meat scraps they were harassed by the wolves. The bears held their ground and did not leave until they had completed their pillaging.

The next morning I was observing the same wolf den when the mother and three cubs were about one-half mile away and moving across a river bar, out of sight. About mid-morning a black, male wolf returned to the den with food in his jaws. He was met by four adults with much tail-wagging and friendly overtures. While the wolves were in a group, a large bear loomed up near the skyline, moving in the general direction of the den. As he came downwind from the wolves he caught the scent of the den, and perhaps the meat also, for he moved toward the den and was about 100 yards away before the wolves discovered him. When the wolves rushed toward the bear, he galloped away but was soon overtaken and surrounded. As the bear dashed at one wolf, another would drive in from behind and the bear would turn quickly to catch it. The wolves avoided his rushes easily. Sometimes a lunge at one wolf was only a feint and the bear would turn and surprise another wolf rushing in from the rear. He would lunge toward the wolf with both paws, not with a slapping movement. After about 10 minutes, the two female wolves withdrew, and within a few minutes the three males had withdrawn also. The bear resumed his travels on a course a little to one side of the den, but the wolves disapproved and again galloped to him. After another 5 minutes of harassment, the wolves returned to the den; the bear retreated the way he had come, and disappeared in a swale one-half mile away. The bear did not touch any wolf, although one escaped the bear's grasp only by the most strenuous efforts. Five wolves had discouraged a lone bear from coming near the den.

Harold Herning reported seeing a grizzly appropriating a calf caribou soon after it was killed by a wolf. Only two of the five wolves present bothered the bear but after being charged by it several times, they retreated. The wolves had an abundance of food and were not near their den, so apparently they felt no strong desire to attack the bear.

In 1940, at a road camp garbage dump, the same female grizzly with the three 2-year-old cubs often met wolves. The cubs frequently chased the wolves but the latter avoided them easily and continued their hunting. One evening the wolves lay down to one side and waited for the bears to leave.

On 22 September 1940 the bear family and the wolves met near the garbage pit. On this occasion the black male chased one of the 2-year-old cubs a short distance, then the cub turned and chased the wolf. Variations of this were repeated several times, both apparently enjoying the game.

On 20 August 1962 a lone wolf was reported attacking a caribou bull which eventually succumbed. (Two other bulls had died from disease about this time which suggests that the bull was an ailing animal.) In the afternoon and evening I watched the wolf feeding on the carcass and carrying off a large piece for caching. The following day the wolf was seen again feeding on the carcass and caching parts of it. In the late afternoon a wolf at the carcass continually watched westward; apparently he was seeing or scenting a bear approaching from that direction, because shortly after the wolf left the carcass a small, dark grizzly appeared from the west, feeding on buffaloberry along the gravel bar. When opposite the carcass, the bear turned abruptly and walked to it. He dragged the remains behind a clump of willows, then carried most of it across a narrow gravel bar to another clump of willow. He carried a large chunk about 100 yards away then returned to the carcass and fed for about 15 minutes before walking away to the west.

On 23 August about 7 a.m. I saw a dark grizzly and the same gray wolf near the carcass. As the bear fed on a piece of neck and ribs, the wolf approached to within 7 or 8 yards. The bear made several short, galloping charges toward the wolf, apparently not hoping to overtake but only to chase it away. Occasionally the bear would follow the wolf and for short spurts break into a gallop. The wolf would keep a little ahead of the bear, 10 yards or less, moving effortlessly and slowly, without excitement, as though only bothered. Once the bear followed the wolf two or three times around a clump of willows about 20 or 25 feet in diameter. After this maneuvering, the wolf picked up a leg bone, moved away 20 yards, and lay down to gnaw on it; the bear resumed feeding on a piece with a few ribs attached. In a few minutes the bear approached the wolf again; the wolf moved away with a caribou leg in his jaws and maneuvered as before, keeping a short distance ahead of the bear. After a time, the wolf dropped his load, but later picked up the piece the bear had been feeding on, carried it a short distance away, and fed. The bear seemed surfeited with meat, or perhaps found the bones too well cleaned, and moved off to feed on buffaloberry. Later he moved far across the flat. The bear had chased half-heartedly and casually, and the wolf, confident of his ability to escape, was not greatly concerned.

In May 1967, a moose carcass near Hogan Creek attracted bears, a wolf, and a wolverine for several days. On 28 May at 3:00 a.m., a lone bear was feeding at the carcass, when, 15 minutes later, a gray wolf trotted down the slope toward the carcass. He passed 50 yards to one side of the carcass, then approached it from below. When the bear saw the wolf 40 yards away, he charged, causing the wolf to retreat some 20 yards. As the bear started back toward the carcass, the wolf followed; the bear turned and charged again. This was repeated at least 25 times before the bear returned to the carcass. The wolf approached to within 10 feet of the bear and after a few token chases of a yard or two, the bear continued feeding as the wolf stood only 7 or 8 feet away. The bear had wearied of discouraging the wolf's approach. If one came upon this scene at this stage, one would assume the wolf and bear to be on the friendliest of terms. The wolf did not attempt to feed on the carcass, and after a few minutes trotted downslope to lie among some scattered spruces. Within an hour the lone bear wandered off, and the wolf, after chasing a wolverine up a tree, came to the carcass and fed undisturbed.

The following morning at 4:45 a.m., the wolf was at the carcass. A mother bear with two 2-year-old cubs appeared, walking rapidly toward the carcass. The wolf remained until the bears were within 25 yards of him, then galloped away lightly, avoiding the charge of the mother bear. He disappeared in some spruces and did not reappear during the 2½ hours that the bear family remained at the carcass, although he was seen in the vicinity later that day.

Young bears sometimes are seen moving away from wolves, perhaps wishing to avoid harassment. One morning, as I watched a wolf working his way diagonally up Primrose Ridge, I saw a small bear coming down the slope ahead of the wolf, perhaps 200 yards away. Apparently the bear caught the scent of the wolf for he raised his nose to test the breeze. Three times he stopped and stood erect on hind legs to watch the wolf. After the third, prolonged look, he dropped to all fours and galloped over a rise. He was well able to take care of himself but preferred keeping his distance. The wolf later noted the bear's trail, followed it a few yards, and then continued on his way.

On another occasion, four wolves were moving leisurely toward the top of Sable Pass. They were scattered, one or two ahead would lie down to wait, while those behind moved here and there nosing about. On the slope ahead, a 3- or 4-year-old grizzly grazed. When he became aware of the wolves, he interrupted his grazing periodically to watch them. After a time, he walked upward and to one side. As he crossed a long snowfield, he glanced toward the wolves several times and disappeared over the horizon. He did not appear alarmed, but as though he preferred to avoid the wolves.

On 4 September 1964 I watched 12 wolves at a rendezvous. In the afternoon a small grizzly appeared near the edge of the sedge flat in which the wolves were resting or moving about. Two pups were playing. The bear was about 150 yards from the nearest wolf when three wolves saw the bear and trotted toward him. In a few moments all 12 wolves were loping toward the bear, and soon he was surrounded. As he faced some of them, others would move in close to his rear, causing him to turn to protect himself. Once, when he began to retreat most of the wolves closed in in a semicircle 3 or 4 yards away from him. He turned and held them at bay, and three circled to his rear. Five black pups soon left the group and later all except one of the adults withdrew. This wolf stood near the bear for 4 or 5 minutes, and when he left, the bear continued on his way. There had been no contact, but the bear probably thought the wolves a nuisance.

If a carcass is involved an adult bear does not retreat from wolves. Once a grizzly appropriated a dead caribou calf even though five wolves were resting nearby. A hungry bear is not to be denied by wolves; he dines with relatively little challenge from that quarter.

Grizzly–Wolverine Relationships

I have learned little about the relationships between the wolverine and the grizzly, having observed a wolverine's reaction to a grizzly on only two occasions. The wolverine probably does not wish to venture near carrion attended by a bear, for he may not be agile enough to escape should the bear attack.

On 21 September 1960 a wolverine that I had been watching for some time, sat up on his haunches when he saw a grizzly about 200 yards away, moving in his direction. He galloped away from the bear, then turned at right angles, traveled a quarter mile, and resumed the line of travel he had been taking when he first saw the bear. He had made a wide detour to avoid the bear, but the bear had given no indication of being aware of the wolverine on this occasion.

On 7 August 1961 I watched a wolverine lope toward a lone bear that was feeding near the river bar, without being aware of the presence of the bear. The wolverine discovered the bear when about 50 yards from it, stopped with a jerk, sat erect, then did an about-face and galloped 100 yards at his fastest pace. Still hurrying, the wolverine climbed the bench above the river and, resuming his original direction, passed well above where the bear was feeding.

In both incidents the wolverine seemed anxious to avoid the bear.

Grizzly–Fox Relationships

The grizzly and the fox often meet at carrion. If a bear is present, the fox may wait patiently for an opportunity to partake.

One spring, a fox and a bear were involved briefly with a cache. A fox dug into a snowfield and secured a food item which he carried 100 yards and recached near a tuft of grass. The robbed cache may have belonged to a bear, because a few hours later one walked to it and fed for 15 minutes on what remained. He then followed the fox's trail and ate its cache. A little later I saw the fox following the bear as they went out of sight over a rise.

Occasionally, a bear may try to dig out a fox's den, but I have seen this only once when two 2-year-old cubs showed an interest. The two cubs spent some time at a den on a knoll, digging haphazardly, with a fox standing a few steps away, watching them and avoiding the half-hearted charges made by one of the cubs. The mother bear, 300 yards away, turned back to check on the tarrying cubs. One came forth and met her 150 yards from the den. The mother turned and started to leave, but the cub moved up the slope to feed, leaving her alone again. The mother again started back to the cub at the fox's den. She was joined by the cub near her, and both walked to the den. The mother left at once, and after some delay, the cubs followed her. The fox had left for another den almost half a mile away, which it approached in a state of excitement as indicated by the tail extended vertically, straight as a ramrod. There may have been fox pups at both dens because this incident occurred in the middle of July when pups are large enough to move from one den to another.

I have seen many fox's dens but only one showed any indication that a bear had tried excavation to get at the young. Dens usually have several entrances so that a bear might have difficulty digging out a fox. A bear

Fig. 58. I have observed numerous fox dens located in choice bear country but rarely have seen any disturbance of them by bears.

had dug into several entrances of a fox's den at Milepost 48, but had not excavated deeply (Fig. 58).

When bears feed in the vicinity of a fox's den, a parent may keep a sharp eye on the bears. One day in early June a pair of bears and two lone bears were digging roots on a river bar. For over an hour I watched a fox sit erect on haunches near her den eyeing the bears, the nearest one being about 150 yards away.

In late July, two bear families grazed all day between 200 and 300 yards from a fox's den. Much of the time one to three foxes could be seen watching the bears. A dozen ground squirrels on the slope between the foxes and the bears also were alert and uttering alarm calls—well they might with two of their most potent enemies in view.

Three different observers have reported seeing a fox play with a grizzly cub. Apparently the play in each case did not involve body contact and the grizzly mothers were indifferent to the play activity.

On 25 September 1963 I watched a fox show special interest in a bear family, for reasons I did not discover. Possibly the bears had fed on meat and the scent lingered. The mother bear was resting on a slope 30 yards above her two yearlings, which were lying where she had nursed them recently. A red fox walked within 10 yards of the cubs and jumped away when one of them sat up. The mother raised her head to look. The bears resumed resting and the fox, after sitting on his haunches a few moments, climbed the slope within 7 or 8 yards of the mother. When she raised her head, he jumped back a yard or two and circled close below her. The mother, perhaps slightly puzzled, walked to her cubs. The fox made a nose inspection of her bed, and departed. Nothing very significant occurred. Animals have their little interests that they must follow up—somewhat like humans.

9
Grizzlies and Birds/Insects

Grizzly–Golden Eagle Relationships

Any bear on the landscape is worthy of at least a brief inspection by an eagle or other animal interested in carrion, for the bear may be at a carcass. I have watched eagles perched on a slope near a bear at a carcass, patiently waiting and hoping for a chance to eat. And I have often watched an eagle circling over a bear, alighting nearby or diving low over him, with no apparent purpose except idle curiosity or casual play. Like a typical neighbor, he is interested in what the neighbors are up to.

On 9 July 1948 I stopped on Sable Pass and, while scanning the country looking for migrating caribou, saw an eagle perched on the point of a yellow bluff. One hundred yards away a bear was making a considerable excavation in his efforts to capture a ground squirrel. After another 10 minutes of digging, the ground squirrel emerged from one of the excavated burrows and was captured after the bear had made four or five jumps after it. As soon as the squirrel was captured, the eagle sailed low over the bear and alit 100 yards up the slope. After eating the squirrel, the bear rambled toward the eagle who took flight with the aid of a few hops when the approaching bear was only about 10 yards away.

The bear walked south until he came to another set of ground squirrel holes. The eagle alit on the slope not far off, the bear dug out a squirrel, and the eagle flew low over him and alit on the slope beyond. Again the bear walked toward the eagle, flushed it, then moved on and excavated a third ground squirrel while the eagle watched from a nearby hummock. The eagle continued following the activities of the bear in this manner and watched him capture six squirrels in six excavations. After this rather phenomenal success at squirrel hunting, the bear turned to grazing on grass and herbs in a green hollow. The eagle had been accompanying the bear for 1½ hours while I watched. He made no attempt to capture any of the squirrels. Why did he stay with the bear? One could imagine that the eagle was comparing the bear's laborious technique in capturing ground squirrels with his own effortless method of gliding low over the country, appearing suddenly over one sharp ridge after another, and sooner or later surprising a squirrel too far from a burrow to escape. (In McKinley National Park the chief food of the eagle is ground squirrel.) (Fig. 59).

On 5 September 1964 I watched a grizzly on Sable Pass digging out a ground squirrel. He was so concerned over the possibility of the squirrel escaping from one of the other exits that he was afraid to dig. He would put his paws in position to pull loose a chunk of sod, then look around to see if the squirrel were escaping, return to digging, but before proceeding, look around a second and even a third time. Three eagles hovered on set wings in a strong wind high over the bear. Later, one of the eagles perched about 25 yards from the bear and another alit about 200 yards up the slope. The third eagle was alternately swooping low over the bear and hovering a short distance above him. After 20 minutes, the bear caught the squirrel deep in the hole and ate it daintily in five or six pieces. The hovering eagle, if he swooped at the right moment, might have captured an escaping ground squirrel, but such an opportunity would be rare because a squirrel trying to escape from a set of holes is usually captured quickly by the bear. However, a photographer in the park reported seeing an eagle capture an escaping ground squirrel after perching near a bear digging for it.

The Golden Eagle and the grizzly hunt ground squirrels and both are attracted to carrion. There is enough for both. Esthetically, their activities add much to the spirit of this wilderness.

Grizzly–Magpie Relationships

Magpies and grizzlies often meet at carrion, a banquet table attractive to all sorts of characters. There is no conflict between these two; the bear takes his share and the magpie is pleased to salvage crumbs that the grizzly considers insignificant.

Occasionally, a magpie is on the scene when a bear excavates a ground squirrel. He sits or hops about while the bear feeds delicately on the squirrel, a small piece at a time. When the bear leaves, the magpie investigates, hopeful that a taste is left. The bear, as he leaves, may see the magpie approach the feeding spot and hurry back to be sure nothing was missed. The always optimistic magpie considers the bear worthy of at least a casual check as he patrols his foraging domain. Sometimes the magpie seems to tease bears for casual amusement. One day, two magpies alit over and over again near two spring cubs, close enough so that a cub twice chased one of the birds. On another occasion I saw a grizzly chase a magpie that had landed where the bear had been resting a few minutes earlier.

Fig. 59. Both golden eagles and grizzlies hunt ground squirrels and are attracted to carrion. There is enough for both.

In farming country we find birds, such as blackbirds and gulls, following the plow to feed on larvae, worms, and insects that have been exposed. This activity has its counterpart in the wilderness. In McKinley National Park I often have observed magpies keeping bears company while they dug roots and examining minutely the freshly turned sod. One day in September, for instance, a mother bear and her two yearlings did considerable digging on a long slope, each bear off by itself, creating scattered black patches of overturned sod here and there. They were attended by four magpies who were searching the turned-over sod, apparently for insect life. Two hours after the bears had left, the magpies were still foraging industriously in the diggings with such silent concentration that one would think they had just made the discovery. The relationship between these two species chiefly benefits the magpies, but I like to imagine that the birds add a little interest to a bear's life.

Grizzly–Raven Relationships

Occasionally a raven has been seen attending a bear digging roots. As the raven forages in the freshly turned sod, he may be feeding only a few feet away from the bear. Ravens occasionally join bears at carrion, as do the Short-billed Gulls.

Like magpies, ravens are a diversion for bears, may be chased halfheartedly at carrion, but usually are ignored.

Grizzly–Insect Relationships

In the high country, bears do not seem to be affected by insects. However, on 15 July 1947, in the woods along the Toklat River where I was watching for wolves, a grizzly's rest was disturbed considerably. I discovered the bear lying in a caribou trail about 40 yards from me. It was a big male that had been climbing a slope and flopped in the trail the moment the lie-down notion struck him. He lay sprawled on his stomach. At intervals he raised his head a few inches and shook it. The mosquitoes were abundant and apparently were bothering him. After a time he became restless, moved down the slope out of my view, and reappeared in some tall willow brush on the edge of the bar. After scratching on a dead snag, he moved across the bar, taking each channel at a gallop, with much splashing and spray. Perhaps other instances when bears seek water are related to pesky insects, but this is usually not evident.

10
Grizzlies and Man

Not surprisingly in an area like McKinley National Park where grizzlies are plentiful and people visit in increasing numbers, interactions between these two species are frequent. Many interactions end amiably, with neither participant suffering unduly. At other times, the people involved may gain a thrill from an imagined charge or being close to a grizzly, but the bear, seemingly untroubled by the encounter, may suffer in the long run. It may be scared away from a choice part of its feeding range, a relatively minor irritation perhaps, but more serious is the effect such an experience may have on predisposing the bear to future encounters that might result in a less innocuous outcome. Those acquainted with bears probably would agree that a tame bear is more dangerous than an unspoiled one in the wilderness.

The encounters we hear about most are those resulting in injury to the people involved, but these cases are a small proportion of all interactions. Herrero (1970) has analyzed cases of human injuries resulting from grizzly encounters, and elsewhere (Murie 1961) I have discussed grizzly relationships with man and recounted a number of incidents in McKinley National Park. Hence I shall consider the subject only briefly here.

In recent years my file on grizzly encounters in McKinley National Park has grown. More and more photographers have taken pictures of grizzlies, and their zeal for "filling the frame" sometimes has resulted in unsettling moments for them. In the spring of 1967 a moose carcass near Savage River attracted several bears, a wolf, and a wolverine over a period of several days. An eager photographer set up his tripod near the carcass hoping for a picture of a mother bear with two 2-year-old cubs that had been feeding on the carcass for several days. This family approached the carcass, saw the man nearby, and all three charged to within 6 feet before his shouting took effect and they all turned and galloped away. The bears did not have the man's scent from their position so perhaps did not realize at first what they were charging. They may

Fig. 60. Photographers, in their zeal for "filling the frame," sometimes experience unsettling moments.

have thought he was a wolf; one had been at the carcass earlier in the day and its scent may have remained in sufficient strength to cause such a mistake. Fortunately, in this and similar incidents, the grizzlies involved were not so accustomed to man that they had lost their usual reaction to him, namely, fleeing, even here when they may have felt their cache was in jeopardy (Fig. 60).

Another incident is worth noting because the bear attack was unprovoked. On 4 August 1961, at 3:50 p.m., an ecologist, Napier Shelton, was near the timberline on a slope of Igloo Mountain taking increment borings, when he heard an ominous growl. About 10 yards away he saw a grizzly coming through the willows and climbed up the tree he was working on which had a 14-inch butt and was 25 feet tall. It was an easy tree to climb because it had a slight downhill slant and rather stout, horizontal limbs that reached almost to the ground. Using the horizontal limbs, the bear was able to paw his way up, and, grabbing Nape's heel, clamped down on the calf of his leg, making deep tooth wounds. Nape hung on but was pulled down a little when the bear slid to the ground; again the bear managed to clamber up, a little higher than the first time, and bit into the thigh of Nape's other leg. All this time Nape was kicking at the bear with his free leg until the animal let go and slid down to the ground. The bear circled the tree two or three times before moving off.

Nape waited in the tree for half an hour before starting down the slope to the road, and it was here that I met him. The bites were deep and one tooth wound had slashed into another. Bleeding was not severe, and he was flown to Fairbanks where the wounds soon healed. The doctor said slyly that he thought he detected a little blueberry juice in one of the wounds!

I had been on Igloo Mountain that same day, a little to one side of the spot where this incident occurred, and I had seen a mother bear with two spring cubs across the creek on the lower slope of Cathedral Mountain, almost opposite the site of the attack. These bears disappeared in a ravine and I surmised that they followed it to the creek and climbed up Igloo Mountain. I had seen the bears just a short time before the attack. Thus it seems likely that it was this mother that happened upon Nape in the willow thicket (Fig. 61).

The incident illustrates the danger of coming suddenly upon a bear, or vice versa. Many bear incidents in wild country result from encounters at close range where both parties are surprised. It is often easier for the bear to attack than to run.

The other situation to be avoided is interposing oneself between a mother and her cub, even if the distance from the mother seems safe. This action is a serious provocation. In fact, any proximity to a family can be dangerous because it is difficult to know just what will pique the mother.

For some reason the public is unafraid of bears. Perhaps this is because real bears resemble so closely harmless Teddy bears. This attitude is justified to a certain extent because bears, on the whole, are rather good-tempered and well-behaved. But the danger lies in their potentiality for causing serious injuries and the uncertainty of their behavior. A half-hearted attack or a casual swipe with a paw can cause a damaging or fatal wound.

When a friend of mine, about to embark into bear country, inquired about the danger of bears, I replied that he had nothing to worry about, that he could travel the wilderness with a light spirit, and that all he needed was faith. The latter, I pointed out, is the chief difficulty. As one gains experience with bears one tends to lose faith, but still if the faith were kept all would be well. The wariest people in the hills are trappers and bear hunters, but after all, they prefer wandering over the hills to crossing a street in modern traffic. The moral is to respect the bear's potential for causing injury and to keep at a respectful distance.

Fig. 61. Napier Shelton, student from Duke University, standing beside a spruce tree where a grizzly attacked him. Shelton climbed the tree and so did the bear, biting him rather severely in both legs. The strong horizontal limbs and slight lean of the tree made it possible for the bear to claw its way up. Picture taken about a week after the encounter.

11
Keeping Grizzlies Wild

This section could be entitled *Bear Management in National Parks*, but I shy away from the word "management" because it has been misused and the less we have of it in national parks, the better. Wildlife managers want to manage everything, just as a forester wants to practice forestry in parks, and engineers want to build more and wider roads.

Whatever management activities we approve should be thought of and undertaken as exceptions. There are, unfortunately, some striking exceptions. Where a fauna is endangered by man's interference, such as in the Everglades, remedial measures are justified. Or when an animal, due to man's activity, is destroying a habitat, as in the case of elk in Yellowstone, control is justified. But again, these adjustments should be regarded as exceptions. The goal is to have minimum manipulation in our parks, to allow, where at all possible, the existing ecological factors to operate naturally. To artificially maintain the picture as first found by Europeans, assuming we know what it was, destroys the significance of the landscape. As an editorial on this point in *Living Wilderness* concluded, "Let us be guardians rather than gardeners."

The report on national parks by the National Academy of Sciences in 1963 summed up what I believe the true objective of national parks should be:

> The Committee recognizes that national parks are not pictures on the wall; they are not museum exhibits in glass cases. They are dynamic biological complexes with self-generating changes. To attempt to maintain them in any fixed condition, past, present, or future, would not only be futile but contrary to nature. Each park should be regarded as a system of interrelated plants, animals, and habitat (an ecosystem) in which evolutionary processes will occur under such human control and guidance as seems necessary to preserve its unique features. Naturalness, the avoidance of artificiality, should be the rule.

This philosophy on parks was often expressed by my highly respected lumberjack and conservationist friend who has spent many summers in McKinley National Park, often in succinct but meaningful "logger language." When a tourist asked him "Where are all the animals?", he replied, "This ain't no zoo, lady." She was given true park service policy in five words. No apologetic hedging or a promise that we would bring animals to the roadside soon. Take Nature as she is, for only then can there be quality experience. McKinley National Park is one of the few places where a sizeable natural population of bears is protected, a unique area where grizzlies and other animals can share a relatively unspoiled land and where people can see them as they ought to be, wild and free. In parks such as Glacier and Yellowstone, grizzlies are more difficult to see in rugged and wooded country, and bears there have been corrupted by exposure to man's refuse. Only recently have steps been taken to return grizzlies to a more natural state. In McKinley National Park we have the chance to avoid some of the serious problems that have led to an association with man that is much too close.

Grizzlies have an uncommon predilection for human foods, whether in the form of garbage or groceries in a cabin. Accessible garbage is a chief cause of bear trouble. Not only does it attract but it continues to hold bears in an area so that they become unafraid and are soon breaking into tents, trailers, or cabins in search of more food. Human contacts follow, and incidents occur where people are harmed, sometimes seriously. The bears become pests instead of remaining interesting wild creatures with natural habits. The usual ending to the story is injury and damage to property, and death to the bear.

When camping in bear country, I have always burned all garbage, including cans to destroy the odors. Taking these precautions has resulted in very little trouble with bears.

If food is stored in cabins, strong, bear-proof shutters should be used to protect the doors and windows. An alternative might be food placed in a cache built on top of four poles. I believe it would be desirable to build a picturesque cache at each of the outlying cabins in the park and store provisions in them instead of in cabins.

In some areas, trouble with bears has been reduced by live-trapping and transporting the animals to distant areas if such are available. To minimize trouble with bears, a combination of all precautions and remedies is needed. In national parks it is undesirable to have any garbage available so that bears will not be attracted to habitations and will not eat such fare, but live in their normal, primitive way.

In the past some researchers have proposed marking grizzly bears and other animals in McKinley National Park to aid in proposed ecological studies. In some studies elsewhere in recent years, grizzlies and other species have been marked with ear tassels for ready identification and have also had radio transmitters attached to them. Some elk in Jackson Hole carry collars of varied hues, moose are ear-tagged, and I have seen trumpeter swans wearing plastic collars. Sensitive people who are sincerely interested in preserving wilderness are opposed to the use of such techniques in an area devoted to esthetics and spiritual values. The observation of tassels in the ears, and the knowledge that the bears have been manhandled systematically, would destroy for many people the wilderness esthetics for an entire region. We might, of course, imagine a conservation situation so critical that such intrusive, harmful techniques would seem necessary. But in the case of the grizzly in McKinley National Park the added information obtainable would not merit the sacrifice of the intangible values for which parks are cherished. In our wilderness parks, research technique should be in harmony with the spirit of wilderness, even though efficiency and convenience may at times be diminished.

It is true that in a highly publicized study in Yellowstone National Park grizzlies carry tassels and radio transmitters. It is also true that when we think of Yellowstone grizzlies, we do not think of wilderness animals, but rather of radios, anesthetized bears, and general manhandling. Surely that study should not set a precedent for McKinley National Park where the grizzly is an outstanding wildlife attraction and the blemish of tagging would be especially disastrous to park esthetics. Although a marking study would make our understanding of grizzly ecology more complete, it is not needed for a sufficiently thorough understanding of the ecology of McKinley grizzlies to enable us to know what is needed for their preservation.

The national park idea is one of the bright spots in our culture. The idealism in the park concept has made every American visiting the national parks feel just a little more worthy. Our generosity to all creatures in the national parks, this reverence for life, is a basic tradition, fundamental to the survival of park idealism. Perpetuation of truly wild grizzlies in McKinley National Park is essential to maintain this tradition (Fig. 62).

Fig. 62. "It would be fitting, I think, if among the last manmade tracks on earth could be found the huge footprints of the great brown bear." (Earl Fleming, 1958).

References

ALTMANN, M. 1956. Moose, *Alces alces,* battles horse in water. *J. Mammal.* 36(1): 145–146.

BERGMAN, S. 1936. Observations on the Kamchatkan bear. *J. Mammal.* 17(2): 115–120.

CHAPMAN, J.A., J.T. Romer and J. Stark. 1955. Ladybird beetles and army cutworm adults as food for grizzly bears in Montana. *Ecology* 36(1): 156–158.

CONLEY, J.D. 1956. Moose vs. bear. *Wyo. Wildl.* 20(9): 37.

CRAIGHEAD, J.J., and F.C. CRAIGHEAD, Jr. 1963. An ecological study of the grizzly bear. Progress report in Quarterly Report Montana Cooperative Wildlife Research Unit 13(3): 20–25.

CRAIGHEAD, J.J., and F.C. CRAIGHEAD, Jr. 1967. Management of bears in Yellowstone National Park. Environ. Res. Inst. and Mont. Coop. Wildl. Res. Unit Rep. 113 p. Unpublished.

ERICKSON, A.W., and L.H. MILLER. 1963. Cub adoption by the brown bear. *J. Mammal.* 44(4): 584–585.

FLEMING, E.J. 1958. Do brown bears attack? *Outdoor Life,* Nov. p. 41.

HERRERO, S. 1970. Human injury inflicted by grizzly bears. *Science* 170: 593–598.

HERRERO, S. (ed.) 1972. Bears—Their Biology and Management. Second International Conf. on Bear Research and Management. IUCN Publ. new series no. 23. IUCN, Morges, Switzerland. 371 p.

KISTCHINSKI, A.A. 1972. Life history of the brown bear (*Ursus arctos* L.) in Northeast Siberia. *In* S. Herrero, ed. Bears—Their Biology and Management. Second International Conf. on Bear Research and Management. IUCN Publ. new series No. 23. IUCN, Morges, Switzerland, 371 p.

LENTFER, J. 1966. Brown-grizzly bear. Work plan segment report. Alaska Dept. Fish and Game. Mimeo. 54 p.

MARTINKA, C.J. 1974. Population characteristics of grizzly bears in Glacier National Park, Montana. *J. Mammal.* 55(1): 21–29.

MERRIAM, C.H. 1918. Review of the grizzly and big brown bears of North America. N. Amer. Fauna No. 41. U.S. Govt. Printing Office, Washington, D.C.

MOOREHEAD, A. 1959. No Room in the Ark. H. Hamilton, London. 227 p.

MUNDY, K.R.D. 1963. Ecology of the Grizzly Bear (*Ursus arctos* L.) in Glacier National Park, British Columbia. M.S. Thesis. Univ. Alberta, Edmonton. 103 p.

MURIE, A. 1937. Some food habits of the black bear. *J. Mammal.* 18(2): 238–240.

MURIE, A. 1944. The Wolves of Mount McKinley. National Park Service, Fauna Series No. 5. U.S. Govt. Printing Office, Washington, D.C. 238 p.

MURIE, A. 1961. A Naturalist in Alaska. The Devin-Adair Co., New York. 302 p.

MURIE, O.J. 1959. Fauna of the Aleutian Islands and Alaska Peninsula. N. Amer. Fauna No. 61. U.S. Govt. Printing Office, Washington, D.C.

PEARSON, A.M. 1972. Population characteristics of the Northern Interior grizzly in the Yukon Territory, Canada. *In* S. Herrero, ed. 1972. Bears—Their Biology and Management. Second International Conf. on Bear Research and Management. IUCN Publ. new series No. 23. IUCN, Morges, Switzerland. 371 p.

PORSILD, A.E. 1953. Edible plants of the Arctic. *Arctic* 6(1): 15–34.

RAUSCH, R.L. 1953. On the status of some arctic mammals. *Arctic* 6(2): 91–148.

SETON, E.T. 1929. Lives of Game Animals. Vol. II, Part 1. Doubleday, Dovan and Co., Inc., New York. 367 p.

SHELDON, C. 1912. Wilderness of the North Pacific Coast Islands. Charles Scribner and Sons, New York.

SHELDON, C. 1930. The Wilderness of Denali. Charles Scribner and Sons, New York. 412 p.

TROYER, W.A. and R.J. HENSEL. 1962. Cannibalism in brown bear. *Anim. Behav*. 10:3–4.

TROYER, W.A., and R.J. HENSEL. 1964. Structure and distribution of a Kodiak bear population. *J. Wildl. Mgmt*. 28:769–772.

Index

☆U.S.GOVERNMENT PRINTING OFFICE: 1981--347-756